NTC POCKET REFERENCES

Dictionary of Chemistry

NTC POCKET REFERENCES

Dictionary of Chemistry

NTC Publishing Group
Lincolnwood, Illinois USA

Library of Congress Cataloging-in-Publication Data

Dictionary of chemistry.
p. cm. -- (NTC pocket references)
Originally published: Oxford England : Helicon Pub. Ltd., c1993.
ISBN 0-8442-0922-8 (alk. paper)
1. Chemistry--Dictionaries. I. Series
QD5.D4984 1996
540'.03--dc20 95-53965
CIP

Manufactured in the United Kingdom.

Editorial director
Michael Upshall

Consultant editor
Thomas G Barrie CChem, MRSC

Project editor
Sara Jenkins-Jones

Text editor
Catherine Thompson

Art editor
Terence Caven

Additional page make-up
Helen Bird

Production
Tony Ballsdon

A

absolute zero the lowest temperature theoretically possible, zero degrees kelvin, equivalent to –273.16°C/–459.67°F, at which molecules are motionless. Although the third law of ◊thermodynamics indicates the impossibility of reaching absolute zero exactly, a temperature within 3×10^{-8} kelvin of it was produced in 1984 by Finnish scientists. Near absolute zero, the physical properties of some materials change substantially; for example, some metals lose their electrical resistance and become superconductive.

acetate common name for ◊ethanoate.

acetic acid common name for ◊ethanoic acid.

acetone common name for ◊propanone.

acetylene common name for ◊ethyne.

acid compound that releases hydrogen ions in the presence of water. The presence of water is essential; acidity is a property of dilute acids. The three most common acids used in the laboratory are sulphuric acid (H_2SO_4), hydrochloric acid (HCl) and nitric acid (HNO_3). These are sometimes referred to as the mineral acids. An acid can also be defined as a proton donor.

$$HCl_{(g)} + aq \leftrightarrow H^+_{(aq)} + Cl^-_{(aq)}$$

The reactions of acids are the reations of the $H^+_{(aq)}$ ion. These are as follows.

with indicators They give a specific colour reaction with indicators; for example, litmus turns red.

with alkalis They react to form a salt and water (neutralization).

$$HCl_{(aq)} + NaOH_{(aq)} \rightarrow NaCl_{(aq)} + H_2O$$

with carbonates With carbonates and hydrogencarbonates, acids form a salt and displace carbon dioxide.

$$HNO_3 + NaHCO_3 \rightarrow NaNO_3 + CO_2 + H_2O$$

with metals Acids react with metals to give off hydrogen and form a salt.

$$Mg + H_2SO_4 \rightarrow MgSO_4 + H_2$$

Acids react with many ◊bases, such as oxides and hydroxides, but the product is not always soluble in water so the reaction soon ceases, as when sulphuric acid reacts with calcium oxide, hydroxide, or carbonate.

Acids can be classified according to their basicity (the number of hydrogen atoms available to react with a base) and degree of ionization (how many of the available hydrogen atoms dissociate in water). Dilute sulphuric acid is classified as a strong (highly ionized), dibasic acid.

Most naturally occurring acids are found as organic compounds, such as the fatty acids R-COOH and sulphonic acids R-SO_3H, where R is an organic molecular structure.

acidic oxide oxide of a ◊non-metal. Acidic oxides are covalent compounds. Those that dissolve in water, such as sulphur dioxide, give acidic solutions.

$$SO_2 + H_2O \leftrightarrow H_2SO_{3(aq)} \leftrightarrow H^+_{(aq)} + HSO^-_{3(aq)}$$

All acidic oxides react with alkalis to form salts.

$$CO_2 + NaOH \rightarrow NaHCO_3$$

acid rain rain with a pH less than 5, thought to be caused principally by the release into the atmosphere of sulphur dioxide (SO_2) and oxides of nitrogen. Sulphur dioxide is formed from the burning of fossil fuels such as coal that contain high quantities of sulphur, and nitrogen oxides are contributed from industrial activities and car exhaust fumes.

Acid rain is linked with damage to and death of forests and lake organisms in Scandinavia, Europe, and eastern North America. It also results in damage to plants and buildings.

acid salt compound formed by the partial neutralization of a dibasic or tribasic ◊acid Although a salt, it contains replaceable hydrogen, so it may undergo the typical reactions of an acid. Examples are sodium hydrogen sulphate ($NaHSO_4$) and acid phosphates.

actinide any of a series of 15 radioactive metallic chemical elements with atomic numbers 89 (actinium) to 103 (lawrencium).

activation energy the energy required to start a chemical reaction. Some elements and compounds will react together merely by bringing them into contact (spontaneous reaction). For others it is necessary to supply energy in order to start the reaction, even if there is ultimately a net output of energy. This initial energy is the activation energy.

The ◊energy of reaction denotes the net change in energy for the reaction as represented by a chemical equation, and does not include the activation energy.

activity series alternative name for ◊reactivity series.

addition polymerization ◊polymerization reaction in which a single monomer gives rise to a single polymer, with no other reaction products.

addition reaction reaction in which the atoms of an element or compound react with a double or triple bond in an organic compound by opening up one of the bonds and becoming attached to it, as when hydrogen chloride reacts with ethene to give chloroethane.

$$CH_2{=}CH_2 + HCl \rightarrow CH_3CH_2Cl$$

An example is the addition of hydrogen atoms to ◊unsaturated compounds in vegetable oils to produce margarine.

adhesive substance that sticks two surfaces together. Natural adhesives include gelatin in its crude industrial form (made from bones, hide fragments and fish offal) and vegetable gums. Synthetic adhesives include thermoplastic and thermosetting resins, which are often stronger than the substances they join; mixtures of epoxy resin and hardener that set by chemical reaction; and elastomeric (stretching) adhesives for flexible joints.

aerated water water that has had air (oxygen) blown through it. Such water supports aquatic life and prevents the growth of bacteria.

aerial oxidation reaction in which air is used to oxidize another substance, as in the contact process for the manufacure of sulphuric acid, and in the ◊souring of wine.

$$2SO_2 + O_2 \leftrightarrow 2SO_3$$

affinity force of attraction (see ◊bond) between chemical elements, which helps to keep them in combination in a molecule. A given element may have a greater affinity for one particular element than for another (for example, hydrogen has a great affinity for chlorine, with which it easily and rapidly combines to form hydrochloric acid, but has little or no affinity for argon).

air see ◊atmosphere.

air pollution contamination of the atmosphere caused by the discharge, accidental or deliberate, of a wide range of toxic substances. Often the amount of the released substance is relatively high in a certain locality, so the harmful effects are more noticeable. The cost of preventing any discharge of pollutants into the air is prohibitive, so attempts are more usually made to reduce gradually the amount of discharge and to disperse this as quickly as possible by using a very tall chimney, or by intermittent release.

alcohol any member of a group of organic chemical compounds characterized by the presence of one or more OH (hydroxyl) groups in the molecule, and which form ◊esters with acids. The main uses of alcohols are as solvents for gums and resins; in lacquers and varnishes; in the making of dyes; for essential oils in perfumery; and for medical substances in pharmacy. Alcohol (ethanol) is produced naturally in the ◊fermentation process and is consumed as part of alcoholic beverages.

Alcohols may be liquids or solids, according to the size and complexity of the molecule. The five simplest alcohols form a series in which the number of carbon and hydrogen atoms increases progressively, each one having an extra CH_2 (methyl) group in the molecule: methanol or wood spirit (methyl alcohol, CH_3OH); ethanol (ethyl alcohol, C_2H_5OH); propanol (propyl alcohol, C_3H_7OH); butanol (butyl alcohol, C_4H_9OH); and pentanol (amyl alcohol, $C_5H_{11}OH$). The lower alcohols are liquids that mix with water; the higher alcohols, such as pentanol, are oily liquids not miscible with water, and the highest are waxy solids – for example, hexadecanol (cetyl alcohol, $C_{16}H_{33}OH$) and melissyl alcohol ($C_{30}H_{61}OH$), which occur in sperm-whale oil and beeswax respectively.

alcoholic solution solution produced when a solute is dissolved in ethanol.

aldehyde any of a group of organic chemical compounds prepared by oxidation of primary alcohols, so that the OH (hydroxyl) group loses its hydrogen to give an oxygen joined by a double bond to a carbon atom (the aldehyde group, –CHO).

The name is made up from ***al***cohol ***dehyd***rogenation, that is, alcohol from which hydrogen has been removed. Aldehydes are usually liquids and include methanal, ethanal, benzaldehyde, formaldehyde, and citral.

aliphatic compound any organic compound that is made up of chains of carbon atoms, rather than rings, as in ◊cyclic compounds. The chains may be linear, as in hexane (C_6H_{14}), or branched, as in 2-propanol (isopropanol) $(CH_3)_2CHOH$.

alkali water-soluble ◊base. The four main alkalis are sodium hydroxide (caustic soda, NaOH); potassium hydroxide (caustic potash, KOH); calcium hydroxide (slaked lime or limewater, $Ca(OH)_2$); and aqueous ammonia ($NH_{3\,(aq)}$). Their solutions all contain the hydroxide ion OH^-, which gives them a characteristic set of properties.

with indicators Alkalis give a specific colour reaction with indicators; for example, litmus turns blue.

with acids They react with acids to form a salt and water (neutralization).

$$KOH + HNO_3 \rightarrow KNO_3 + H_2O$$

$$OH^- + H^+ \rightarrow H_2O$$

with ammonium salts Alkalis displace ammonia gas from ammonium salts.

$$NH_4Cl + NaOH \rightarrow NaCl + NH_3 + H_2O$$

$$NH^+_{4(s)} + OH^-_{(aq)} \rightarrow NH_{3\,(g)} + H_2O$$

with soluble salts Alkalis precipitate the insoluble hydroxides of most metals from soluble salts.

$$FeCl_2 + 2NaOH \rightarrow Fe(OH)_2 + 2NaCl$$

$$Fe^{2+}_{(aq)} + 2OH^-_{(aq)} \rightarrow Fe(OH)_{2\,(s)}$$

alkali metal any of a group of six metallic elements with similar bonding properties: lithium, sodium, potassium, rubidium, caesium, and francium. They form a linked group (group I) in the ◊periodic table of the elements. They are univalent and of very low density (lithium, sodium, and potassium float on water); in general they are reactive, soft, low-melting-point metals. Because of their reactivity they are only found as compounds in nature, and are used as chemical reactants rather than as structural metals.

alkaline-earth metal any of a group of six metallic elements with similar bonding properties: beryllium, magnesium, calcium, strontium, barium, and radium. They form a linked group (group II in the ◊periodic table. They are strongly basic, bivalent, and occur in nature only as compounds.

alkane member of the group of ◊hydrocarbons having the general formula C_nH_{2n+2} (common name ***paraffins***). Lighter alkanes are colourless gases (for example, methane, ethane, propane, butane); in nature they are found dissolved in petroleum. Heavier ones are liquids or solids. As they contain only single ◊covalent bonds, they are said to be saturated.

Their principle reactions are combustion and bromination.

$$CH_4 + 2O_2 \rightarrow CO_2 + H_2O$$

$$C_2H_6 + Br_2 \xrightarrow{200°C} C_2H_5Br + HBr$$

alkene member of the group of ◊hydrocarbons having the general formula C_nH_{2n}, (commonly known as ***olefins***). Lighter alkenes, such as ethene (CH_2=CH_2) and propene (CH_3CH=CH_2), are gases, obtained from the ◊cracking of oil fractions. They are unsaturated compounds, characterized by one or more double bonds between adjacent carbon atoms. They react by addition, and many useful compounds (such as poly(ethene) and bromoethane) are made from them.

alkane

name	*molecular formula*	*structural formula*
methane	CH_4	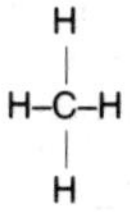
uses: domestic fuel (natural gas)		
ethane	C_2H_6	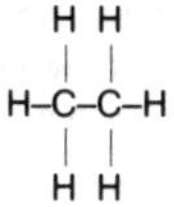
uses: industrial fuel and chemical feedstock		
propane	C_3H_8	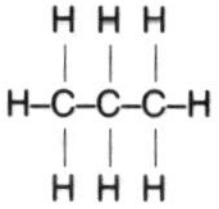
uses: bottled gas (camping gas)		
butane	C_4H_{10}	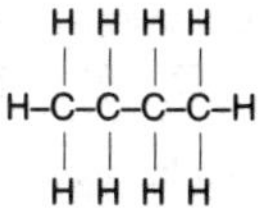
uses: bottled gas (lighter fuel, camping gas)		

alkyne member of the group of ◊hydrocarbons with the general formula C_nH_{2n-2} (commonly known as ***acetylenes***). They are unsaturated compounds, characterized by one or more triple bonds between adjacent carbon atoms. Lighter alkynes are gases (for example, ◊ethyne); heavier ones are liquids or solids.

allotropy the property whereby certain elements exist in different forms (allotropes) in the same physical state. The allotropes of carbon are diamond and graphite; those of ◊sulphur are rhombic and monoclinic sulphur. These have different crystal structures when solid, as do the the white and grey forms of tin. Oxygen also exists in two different forms: 'normal' oxygen (O_2) and ozone (O_3), which have different molecular configurations.

alloy metal blended with some other metallic or non-metallic substance to give it special qualities, such as resistance to corrosion, greater hardness, or tensile strength. Useful alloys include bronze, brass, cupronickel, duralumin, German silver, gunmetal, pewter, solder, steel, and stainless steel. The most recent alloys include the superplastics, alloys that can stretch 100% at specific temperatures, permitting, for example, their injection into moulds as easily as plastic.

Among the oldest alloys is bronze, whose widespread use ushered in the Bronze Age. Complex alloys are now widespread, for example in dentistry, where a cheaper alternative to gold is made of chromium, cobalt, molybdenum, and titanium.

alpha particle positively charged, high-energy particle emitted from the nucleus of a radioactive ◊atom. It is one of the products of the spontaneous disintegration of radioactive elements such as radium and thorium, and is identical with the nucleus of a helium atom – that is, it consists of two protons and two neutrons. The process of emission, ***alpha decay***, transforms one element into another, decreasing the proton number by two and the nucleon number by four. See ◊radioactivity.

Because of their large mass alpha particles have a short range of only a few centimetres in air, and can be stopped by a sheet of paper. They are capable of damaging living cells.

alum any double sulphate of a monovalent metal or radical (such as sodium, potassium, or ammonium) with a trivalent metal (such as aluminium or iron). The commonest alum is the double sulphate of potassium and aluminium, $KAl(SO_4)_2.12H_2O$, a white crystalline powder that is readily soluble in water. Alums are used in papermaking and to fix dye in textiles.

alumina Al_2O_3 or ***corundum*** oxide of aluminium that is widely distributed in clays, slates, and shales. It is formed by the decomposition of the feldspars in granite and used as an abrasive.

Typically it is a white powder, soluble in most strong acids or caustic alkalis, but not in water. Impure alumina is called 'emery'. Rubies and sapphires are corundum gemstones.

aluminium light-weight, silver-white, ductile and malleable, metallic element, symbol Al, atomic number 13, relative atomic mass 26.9815. It is the third most abundant element (and the most abundant metal) in the Earth's crust, of which it makes up about 8.1% by mass. It oxidizes rapidly, the layer of oxide on its surface making it highly resistant to tarnish, and is an excellent conductor of electricity. In its pure state it is a weak metal, but when combined with elements such as copper,

aluminium

the extraction of aluminium from bauxite by electrolysis

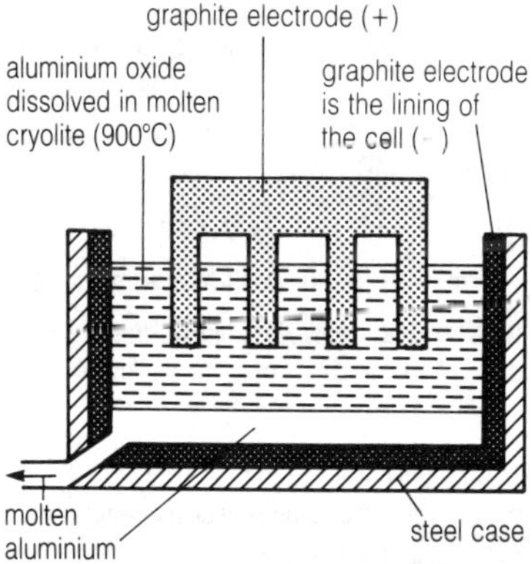

silicon, or magnesium it forms alloys of great strength. In nature it is found only in its combined state in many minerals, and it is prepared commercially from the ore bauxite. In the USA the original name suggested by the scientist Humphry Davy, 'aluminum', is retained.

The pure metal was not readily obtained until the middle of the 19th century. Because of its light weight (specific gravity 2.70) it is widely used in the shipbuilding and aircraft industries. Consumer uses include food and beverage packaging, foil, outdoor furniture, and homebuilding materials. It is also much used in steel-cored overhead cables and for canning uranium slugs for nuclear reactors. It is an essential constituent in some magnetic materials and, as a good conductor of electricity, is used as foil in electrical capacitors. A plastic form of aluminium, developed 1976, which moulds to any shape and extends to several times its original length, has uses in electronics, cars, building construction, and so on.

The metal is extracted from purified ◊bauxite (aluminium oxide, Al_2O_3) dissolved in molten cryolite at 900°C by electrolysis in a Hall cell. The reactions that occur at each electrode are as follows:

negative electrode: $2Al^{3+} + 6e^- \rightarrow 2Al$

positive electrode: $3O^{2-} - 6e^- \rightarrow 1\frac{1}{2}O_2$

The oxygen reacts with the carbon anodes, which must be replaced at intervals.

aluminium chloride $AlCl_3$ white solid made by direct combination of aluminium and chlorine.

$$2Al + 3Cl_2 \rightarrow 2AlCl_3$$

The anhydrous form is a typical covalent compound (see ◊covalency).

aluminium hydroxide $Al(OH)_3$ gelatinous ◊precipitate formed when a small amount of alkali solution is added to a solution of an aluminium salt.

$$Al^{3+}_{(aq)} + 3OH_{(aq)} \rightarrow Al(OH)_{3\,(s)}$$

It is an ◊amphoteric compound as it readily reacts with both acids and alkalis.

aluminium oxide or ◊*alumina* Al_2O_3 white solid formed by heating aluminium hydroxide. It is an ◊amphoteric oxide (reacting with both acids and alkalis) and is used as a refractory (furnace lining) and in column ◊chromatography.

amalgam any alloy of mercury with other metals. Most metals will form amalgams, except iron and platinum. Amalgam is used in dentistry for filling teeth, and usually contains copper, silver, and zinc as the main alloying ingredients. This amalgam is pliable when first mixed and then sets hard, but the mercury leaches out and may cause a type of heavy-metal poisoning.

Amalgamation, the process of forming an amalgam, is a technique sometimes used to extract gold and silver from their ores. The ores are treated with mercury, which combines with the precious metals.

amino acid water-soluble organic molecule, mainly composed of carbon, oxygen, hydrogen, and nitrogen, containing both a basic amine group ($-NH_2$) and an acidic carboxyl group (–COOH). When two or more amino acids are joined together, they are known as ◊peptides; ◊proteins are made up of interacting polypeptides (peptide chains consisting of more than three amino acids) and are folded or twisted in characteristic shapes.

Many different proteins are found in the cells of living organisms, but they are all made up of the same 20 amino acids, joined together in varying combinations (although other types of amino acid do occur infrequently in nature). Eight of these, the *essential amino acids*, cannot be synthesized by humans and must be obtained from the diet. Children need a further two amino acids that are not essential for adults. Other animals also need some preformed amino acids in their diet, but green plants can manufacture all the amino acids they need from simpler molecules, relying on energy from the sun and minerals (including nitrates) from the soil.

ammonia NH_3 colourless, pungent-smelling gas, lighter than air and very soluble in water. It is made on an industrial scale by the ◊Haber process, and used mainly to produce nitrogenous fertilizers, some explosives, and nitric acid.

The gas has several typical reactions.

with metal oxides It will reduce some metal oxides when heated.

$$3CuO + 2NH_3 \rightarrow 3Cu + N_2 + 3H_2O$$

with water It dissolves readily in water to give an alkaline solution. Because ammonia is a weak base, its hydroxide-ion concentration is not high (pH 10).

$$NH_3 + H_2O \leftrightarrow NH_4^+ + OH^-$$

with indicators The gas turns moist litmus blue; this is used as a test for the presence of ammonia.

with hydrogen chloride Ammonia gas combines with hydrogen chloride gas to form white clouds of ammonium chloride.

$$NH_{3\,(g)} + HCl_{(g)} \leftrightarrow NH_4Cl_{(s)}$$

ammoniacal solution solution produced by dissolving a solute in aqueous ammonia.

ammonium NH_4^+ ion formed when ammonia accepts a proton (H^+) from an acid. It is the only positive ion that is not a metal and that forms a series of salts. When an ammonium salt is heated with an alkali it produces ammonia gas.

$$NH_4^+ + OH^- \rightarrow NH_3 + H_2O$$

This reaction is used to detect the presence of the ion. Ammonium is a useful source of nitrogen for plants and is present in many fertilizer preparations.

ammonium carbonate $(NH_4)_2CO_3$ white, crystalline solid that readily sublimes at room temperature into its constituent gases (ammonia, carbon dioxide, and water). It was formerly used in ◊smelling salts.

ammonium chloride or ***sal ammoniac*** NH_4Cl volatile salt that forms white crystals around volcanic craters. It is prepared synthetically for use in 'dry-cell' batteries, fertilizers, and dyes.

ammonium nitrate NH_4NO_3 colourless, crystalline solid, prepared by ◊neutralization of nitric acid with ammonia; the salt is crystallized from the solution. It sublimes on heating. Ammonium nitrate is used in large quantities as a fertilizer and in making explosives.

amphoteric term applied to elements that show the properties of both ◊metals and ◊non-metals. For example, aluminium and zinc react with both acids and alkalis.

$$Zn + 2HCl \rightarrow ZnCl_2 + H_2$$
$$Zn + 2NaOH \rightarrow Na_2ZnO_2 + H_2$$

The term also applies to the oxides and hydroxides of these elements, which also react with both acids and alkalis.

Amino acids and proteins are also amphoteric compounds, as they contain a basic (amino, $-NH_2$) and an acidic (carboxyl, –COOH) group. Thus in an acid solution the amino acid acts as a base, and in an acidic solution it acts as an acid.

analysis the determination of the composition or properties of substances; see ◊analytical chemistry.

analytical chemistry branch of chemistry that deals with the determination of the chemical composition of substances.

Qualitative analysis determines the elements or compounds in a given sample, without necessarily finding their concentrations, using methods such as ◊chromatography and spectroscopy.

Quantitative analysis determines exact composition in terms of concentration, using such techniques as ◊titration (volumetric analysis) and weighing (gravimetric analysis).

anhydride compound obtained by the removal of water from another compound, usually an acid. For example, sulphur trioxide (SO_3) is the anhydride of sulphuric acid (H_2SO_4). For monobasic acids, such as organic fatty acids, the formation of an anhydride involves the loss of a molecule of water from two molecules of acid.

anhydrite naturally occurring anhydrous calcium sulphate ($CaSO_4$). It is used commercially for the manufacture of plaster of Paris and builders' plaster.

anhydrous term describing a substance from which all water has been removed.

If the water of crystallization is removed from blue crystals of copper(II) sulphate ($CuSO_4$), a white powder (anhydrous copper) sulphate results). Liquids from which all traces of water have been removed are also described as anhydrous.

anion ion carrying a negative charge. An electrolyte, such as the salt zinc chloride ($ZnCl_2$), is dissociated in aqueous solution or in the molten state into doubly-charged Zn^{2+} zinc ◊cations and singly-charged Cl^- anions. During electrolysis, the zinc cations flow to the cathode (to become discharged and liberate zinc metal) and the chloride anions flow to the anode (to liberate chlorine gas).

anode the positive electrode towards which negative particles (anions or electrons) move within an electrolytic cell.

anodizing process that increases the resistance to ◊corrosion of a metal, such as aluminium, by building up a protective oxide layer on the surface. The natural corrosion resistance of aluminium is provided by a thin film of aluminium oxide; anodizing increases the thickness of this film and thus the corrosion protection.

It is so called because the metal becomes the ◊anode in an electrolytic bath containing a solution of, for example, sulphuric or chromic acid as the electrolyte. During ◊electrolysis oxygen is produced at the anode, where it combines with the metal to form an oxide film.

anomalous expansion of water the expansion of water as it is cooled from 4°C to 0°C. This behaviour is unusual, because most substances contract when they are cooled. It means that water has a greater density at 4°C than at 0°C. Hence ice floats on water, and the water at the bottom of a pool in winter is warmer than at the surface. As a result large lakes freeze slowly in winter and aquatic life is more likely to survive.

antacid weak base taken medicinally to counter stomach acidity, usually as a powder or an emulsion. Substances used in antacids include sodium bicarbonate (sodium hydrogencarbonate), magnesium hydroxide, magnesium and calcium carbonates, and aluminium hydroxide. ◊Milk of magnesia is a commercial antacid that contains magnesium oxide suspended in water.

antifreeze substance added to a water-cooling system (for example, that of a car) to prevent it freezing in cold weather. The most common types of antifreeze contain the chemical ethylene ◊glycol ($HOCH_2CH_2OH$), an organic alcohol with a freezing point of about –15°C/5°F.

The addition of this chemical depresses the freezing point of water significantly. A solution containing 33.5% by volume of ethylene glycol will not freeze until about –20°C/–4°F. A 50% solution will not freeze until –35°C/–31°F.

antiseptic any substance that kills or inhibits the growth of microorganisms. The use of antiseptics was pioneered by Joseph ◊Lister. He used carbolic acid (◊phenol), which is a weak antiseptic; substances such as TCP are derived from this.

aqueous solution solution in which the solvent is water.

argon (Greek *argos* 'idle') colourless, odourless, non-metallic, gaseous element, symbol Ar, atomic number 18, relative atomic mass 39.948. It is grouped with the ◊inert gases, since it was long believed not to react with other substances, but observations now indicate that it can be made to combine with boron fluoride to form compounds. It constitutes almost 1% of the Earth's atmosphere, and was discovered by the British chemists John Rayleigh and William Ramsay after all oxygen and nitrogen had been removed chemically from a sample of air. It is used in electric light bulbs and radio tubes.

aromatic compound any organic chemical that incorporates a ◊benzene ring in its structure (see also ◊cyclic compounds). Aromatic compounds undergo ◊substitution reactions.

arsenic brittle, greyish-white, semi-metallic element (a metalloid), symbol As, atomic number 33, relative atomic mass 74.92. It occurs in many ores and occasionally in its elemental state, and is widely distributed, being present in minute quantities in the soil, the sea, and the human body. In larger quantities, it is poisonous. The chief source of arsenic compounds is as a by-product from metallurgical processes. It is used in making semiconductors, alloys, and solders.

As it is a cumulative poison, its presence in food and drugs is very dangerous. The symptoms of arsenic poisoning are vomiting, diarrhoea, tingling and possibly numbness in the limbs, and collapse.

artificial radioactivity radioactivity arising from human-made radioisotopes (radioactive isotopes or elements that are formed when elements are bombarded with subatomic particles – protons, neutrons, or electrons – or small nuclei).

ascorbic acid or ***vitamin C*** relatively simple organic acid found in fresh fruits and vegetables. It is soluble in water and destroyed by prolonged boiling, so soaking or overcooking of vegetables reduces their vitamin C content. Lack of ascorbic acid results in scurvy.

asphalt semi-solid brown or black ◊bitumen, used in the construction industry. Asphalt is mixed with rock chips to form paving material, and the purer varieties are used for insulating material and for waterproofing masonry. It can be produced artificially by the distillation of ◊petroleum.

assaying determination of the quantity of a given substance present in a sample. Usually it refers to determining the purity of precious metals.

The assay may be carried out by 'wet' methods, when the sample is wholly or partially dissolved in some reagent (often an acid), or by 'dry' or 'fire' methods, in which the compounds present in the sample are combined with other substances.

astatine (Greek *astatos* 'unstable') non-metallic, radioactive element, symbol At, atomic number 85, relative atomic mass 210. It is a member of the ◊halogen group, and is very rare in nature. Astatine is highly unstable, with many isotopes; the longest lived has a half-life of about eight hours.

atmosphere the mixture of gases that surrounds the Earth, prevented from escaping by the pull of the Earth's gravity. Atmospheric pressure decreases with height in the atmosphere. In its lowest layer, the atmosphere consists of nitrogen (78%) and oxygen (21%), both in molecular form (N_2 and O_2). The other 1% is largely argon, with very small quantities of other gases, including water vapour and carbon dioxide. The atmosphere plays a major part in the various cycles of nature (the ◊water cycle, ◊carbon cycle, and ◊nitrogen cycle). It is the principle industrial source of nitrogen, oxygen, and argon, which are obtained by fractional distillation of liquid air.

atom the smallest unit of matter that can take part in a chemical reaction, and which cannot be broken down chemically into anything simpler. An atom is made up of protons and neutrons in a central nucleus surrounded by electrons (see ◊atomic structure). The atoms of the various elements differ in atomic number, relative atomic mass, and chemical behaviour. There are 109 different types of atom, corresponding with the 109 known elements as listed in the ◊periodic table of the elements.

Atoms are much too small to be seen even by the microscope (the largest, caesium, has a diameter of 0.0000005 mm/0.00000002 in), and they are in constant motion. Belief in the existence of atoms dates back to the ancient Greek natural philosophers. The first scientist to gather evidence for the existence of atoms was John Dalton, in the 19th century, who believed that every atom was a complete unbreakable entity. Ernest Rutherford showed by experiment that an atom consists of a nucleus surrounded by negatively charged particles called electrons.

atom, electronic structure the arrangement of electrons around the nucleus of an atom, in distinct energy levels, also called orbitals or shells (see ◊orbital, atomic). These shells can be regarded as a series of concentric spheres, each of which can contain a certain maximum number of electrons; the noble gases have an arrangement in which every shell contains this number (see ◊noble gas structure). The energy levels are usually numbered beginning with the shell nearest to the nucleus. The outermost shell is known as the ◊valency shell as it contains the valence electrons.

atom, electronic structure

sodium 2.8.1 sulphur 2.8.6

The atomic number of an element indicates the number of electrons in a neutral atom. From this it is possible to deduce its electronic structure. For example, sodium has atomic number 11 ($Z = 11$) and its electronic arrangement (configuration) is two electrons in the first energy level, eight electrons in the second energy level and one electron in the third energy level—generally written as 2.8.1. Similarly for sulphur ($Z = 16$), the electron arrangement will be 2.8.6. The electronic structure dictates whether two elements will combine by ionic or covalent bonding (see ◊bond) or not at all.

atomicity the number of atoms of an ◊element that combine together to form a molecule. A molecule of oxygen (O_2) has atomicity 2; sulphur (S_8) has atomicity 8.

atomic mass unit or ***dalton*** unit (symbol amu or u) of mass that is used to measure the relative mass of atoms and molecules. It is equal to one-twelfth of the mass of a carbon-12 atom, which is equivalent to the mass of a proton or 1.66×10^{-27} kg. The ◊relative atomic mass of an atom has no units; thus oxygen-16 has an atomic mass of 16 daltons, but a relative atomic mass of 16.

atomic number or ***proton number*** the number (symbol Z) of protons in the nucleus of an atom. It is equal to the positive charge on the nucleus. The 109 elements are numbered 1 (hydrogen) to 109 (unnilennium) in the ◊periodic table of elements. See also ◊nuclear notation.

atomic size or ***atomic radius*** the size of an atom expressed as the radius in ◊ångströms or other units of length. The sodium atom has an atomic radius of 1.57 ångströms (1.57×10^{-8} cm). For metals, the size of the atom is always greater than the size of its ion. For non-metals the reverse is true.

atomic structure the internal structure of an ◊atom. The core of the atom is the ***nucleus***, a particle only one ten-thousandth the diameter of the atom itself. The simplest nucleus, that of hydrogen, comprises a single positively charged particle, the ***proton***. Nuclei of other elements contain more protons and additional particles of about the same mass as the proton but with no electrical charge, ***neutrons***. Each element has its own characteristic nucleus with a unique number of protons, the atomic number. The number of neutrons may vary. Where atoms of a single element have different numbers of neutrons, they are called ◊isotopes. Although some isotopes tend to be unstable and exhibit ◊radioactivity, they all have identical chemical properties.

The nucleus is surrounded by a number of ***electrons***, each of which has a negative charge equal to the positive charge on a proton, but which weighs only 1/1839 times as much. For a neutral atom, the nucleus is surrounded by the same number of electrons as it contains protons. The chemical properties of an element are determined by the ease with which its atoms can gain or lose electrons. This is dependent on both the number of electrons associated with the nucleus and the force exerted on them by its positive charge.

Atoms are held together by the electrical forces of attraction between each negative electron and the positive protons within the nucleus. The latter repel one another with relatively enormous forces; a nucleus holds together only because other forces, not of a simple electrical character, attract the protons and neutrons to one another. These additional forces act only so long as the protons and neutrons are virtually in contact with one another. If, therefore, a fragment of a complex nucleus, containing some protons, becomes only slightly loosened from the main group of neutrons and protons, the strong natural repulsion between the protons will cause this fragment to fly apart from the rest of the nucleus at high speed. It is by such fragmentation of atomic nuclei (◊nuclear fission) that nuclear energy is released.

atomic weight another name for ◊relative atomic mass.

Avogadro's hypothesis law stating that equal volumes of all gases, when at the same temperature and pressure, have the same numbers of molecules. This law was first propounded by Amadeo Avogadro.

Avogadro's number or ***Avogadro's constant*** the number of carbon atoms in 12 g of the carbon-12 isotope (6.022045×10^{23}). The relative atomic mass of any element, expressed in grams, contains this number of atoms. It is named after Amadeo Avogadro.

B

Bakelite the first synthetic ◊plastic, created by Leo Baekeland in 1909. Bakelite is hard, tough, and heatproof, and is used as an electrical insulator. It is made by the reaction of phenol with formaldehyde, producing a powdery resin that sets solid when heated (it is a ◊thermosetting plastic). Objects are made by subjecting the resin to compression moulding (simultaneous heat and pressure in a mould). It is often used for electrical fittings.

baking powder mixture of ◊bicarbonate of soda (sodium hydrogencarbonate, $NaHCO_3$) and solid tartaric acid, used in cooking as a raising agent. When added to flour, the presence of water and heat causes carbon dioxide to be released, which makes the dough rise.

barium (Greek *barytes* 'heavy') soft, silver-white, metallic element, symbol Ba, atomic number 56, relative atomic mass 137.33. It is one of the alkaline-earth metals, found in nature as barium carbonate and barium sulphate. As the sulphate it is used in medicine; taken in suspension (a 'barium meal'), its progress is followed by using X-rays to reveal abnormalities of the alimentary canal. Barium is also used in alloys, pigments, and safety matches and, with strontium, forms the emissive surface in cathode-ray tubes. It was first discovered in barytes or heavy spar.

barium chloride $BaCl_2$ white, crystalline solid, used in aqueous solution to test for the presence of the sulphate ion. When a solution of barium chloride is mixed with a solution of a sulphate, a white precipitate of barium sulphate is produced.

$$Ba^{2+}_{(aq)} + SO^{2-}_{4\,(aq)} \rightarrow BaSO_{4\,(s)}$$

barium nitrate $Ba(NO_3)_2$ compound that, like barium chloride, can be used to test for the presence of the sulphate ion as it readily dissolves in water. When sulphate is added, a white precipitate of barium sulphate is formed.

base substance that accepts protons. Bases can contain negative ions such as the hydroxide ion (OH^-), which is the strongest base, or be molecules such as ammonia (NH_3). Ammonia is a weak base, as only some of its molecules accept protons.

$$OH^- + H^+_{(aq)} \rightarrow H_2O_{(l)}$$

$$NH_3 + H_2O \leftrightarrow NH_4^+ + OH^-$$

Bases that dissolve in water are called ◊alkalis.

Inorganic bases are usually oxides or hydroxides of metals, which react with dilute acids to form a salt and water. Many carbonates also react with dilute acids, additionally giving off carbon dioxide.

basicity the number of replaceable hydrogen atoms in an acid. Nitric acid (HNO_3) is monobasic, sulphuric acid (H_2SO_4) is dibasic, and phosphoric acid (H_3PO_4) is tribasic.

basic oxide compound formed by a metal and oxygen, containing the O^{2-} ion. If, like sodium oxide (Na_2O), it is soluble, it forms the metal hydroxide when added to water.

$$Na_2O_{(s)} + H_2O_{(l)} \rightarrow 2NaOH_{(aq)}$$

All basic oxides react with acids to form a salt and water.

$$CaO + 2HNO_3 \rightarrow Ca(NO_3)_2 + H_2O$$

basic-oxygen process the most widely used method of steelmaking, involving the blasting of oxygen at supersonic speed into molten pig iron.

Pig iron from a blast furnace, together with steel scrap, is poured into a converter, and a jet of oxygen is then projected into the mixture. The excess carbon in the mix and other impurities quickly burn out or form a slag, and the converter is emptied by tilting. It takes only about 45 minutes to refine 350 tonnes/400 tons of steel. The basic-oxygen process was developed in 1948 at a steelworks near the Austrian towns of Linz and Donawitz.

basic-oxygen process

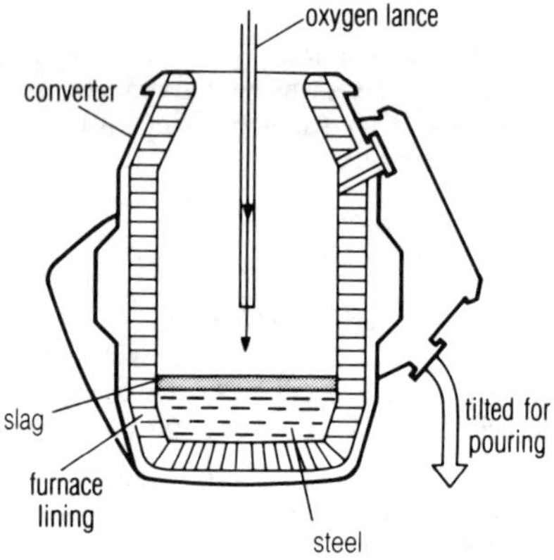

battery any energy storage device allowing release of electricity on demand. A battery is made up of one or more cells, each containing two conducting electrodes (one positive, one negative) immersed in an ◊electrolyte, in a container. When an outside connection (such as through a light bulb) is made between the electrodes, a current flows through the circuit, and chemical reactions releasing energy take place within the cells.

Primary-cell batteries are disposable; secondary-cell batteries are rechargeable. The common ***dry cell*** is a primary-cell battery that consists of a central carbon electrode immersed in a paste of manganese dioxide and ammonium chloride as the electrolyte. The zinc casing forms the other electrode. It is dangerous to try to recharge a primary-cell battery.

The introduction of rechargeable nickel–cadmium batteries has revolutionized portable electronic newsgathering (sound recording, video) and information processing (computing). These batteries offer a stable, short-term source of power free of noise and other hazards associated with mains electricity.

battery

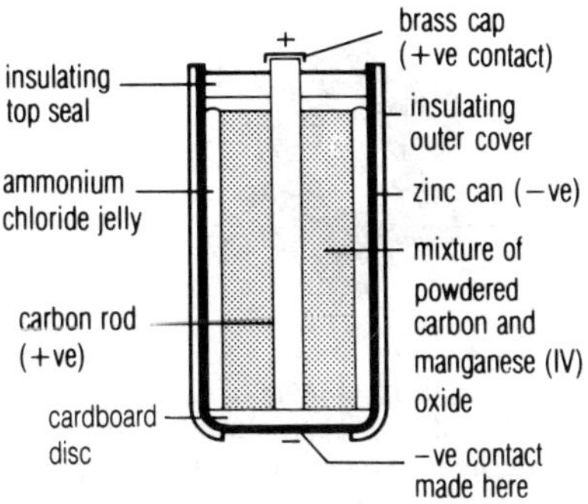

The lead-acid ***car battery*** is a secondary-cell battery, or accumulator. The car's generator continually recharges the battery. It consists of sets of lead (positive) and lead peroxide (negative) plates in an electrolyte of sulphuric acid.

battery acid ◊sulphuric acid of 70% concentration used in lead-cell batteries (as in motor vehicles).

bauxite the principal ore of ◊aluminium, consisting of a mixture of hydrated aluminium oxides and hydroxides, generally contaminated with compounds of iron, which give it a red colour. Chief producers of bauxite are Australia, Guinea, Jamaica, Russia, Kazakhstan, Surinam, and Brazil.

benzene C_6H_6 clear liquid hydrocarbon of characteristic odour, occurring in coal tar. It is used as a solvent and in the synthesis of many chemicals.

The benzene molecule consists of a ring of six carbon atoms, all of which are in a single plane, and it is one of the simplest ◊cyclic compounds. Benzene is the simplest of a class of compounds collectively known as ***aromatic compounds***. Some are considered carcinogenic.

beryllium hard, light-weight, silver-white, metallic element, symbol Be, atomic number 4, relative atomic mass 9.012. It is one of the ◊alka-

benzene

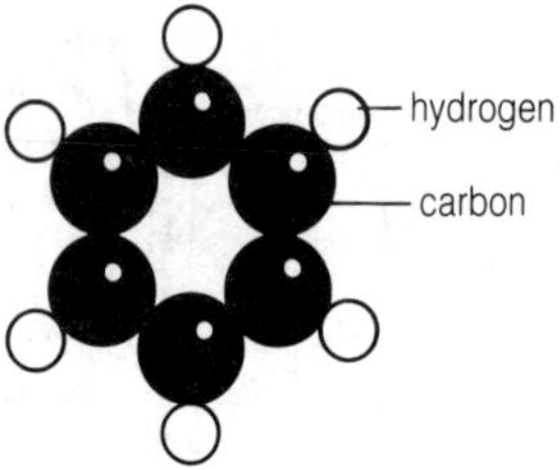

line-earth metals, with chemical properties similar to those of magnesium; in nature it is found only in combination with other elements. It is used to make sturdy, light alloys and to control the speed of neutrons in nuclear reactors. It was discovered in 1798 by French chemist Louis-Nicolas Vauquelin (1763–1829).

beta particle electron ejected with great velocity from a radioactive atom that is undergoing spontaneous disintegration. Beta particles do not exist in the nucleus but are created on disintegration, ***beta decay***, when a neutron converts to a proton to emit an electron. The process transforms one element into another, increasing the proton number by one while the nucleon number remains the same.

Beta particles are more penetrating than ◊alpha particles, but less so than ◊gamma radiation; they can travel several metres in air, but are stopped by 2–3 mm of aluminium. They are less strongly ionizing than alpha particles and, like cathode rays, are easily deflected by magnetic and electric fields.

bicarbonate common name for ◊hydrogencarbonate.

bicarbonate of soda $NaHCO_3$ (technical name ***sodium hydrogencarbonate***) white crystalline solid that neutralizes acids and is used in medicine to treat acid indigestion. It is also used in baking powders and fizzy drinks.

biochemistry science concerned with the chemistry of living organisms: the structure and reactions of proteins such as enzymes, nucleic acids, carbohydrates, and lipids.

The study of biochemistry has increased our knowledge of how animals and plants react with their environment, for example, in creating and storing energy by photosynthesis, taking in food and releasing waste products, and passing on their characteristics through their genes. It plays a part in many areas of research, including medicine and agriculture.

biodegradable capable of being broken down by living organisms, principally bacteria and fungi. Biodegradable substances, such as food and sewage, can therefore be rendered harmless by natural processes. The process of decay leads to compaction and liquefaction, and to the release of nutrients that are then recycled by the ecosystem. Non-biodegradable substances, such as glass, heavy metals, and most types of plastic, present major problems of disposal.

bismuth hard, brittle, pinkish-white, metallic element, symbol Bi, atomic number 83, relative atomic mass 208.98. It is the last of the stable elements; all from atomic number 84 up are radioactive. Bismuth occurs in ores and occasionally as a free metal. It is a poor conductor of heat and electricity, and is used in alloys of low melting point and in medical compounds to soothe gastric ulcers.

bitumen an impure mixture of hydrocarbons, including such deposits as petroleum, asphalt, and natural gas, although sometimes the term is restricted to a soft kind of pitch resembling asphalt.

Solid bitumen may have arisen as a residue from the evaporation of petroleum. If evaporation took place from a pool or lake of petroleum, the residue might form a pitch or asphalt lake, such as Pitch Lake in Trinidad. Bitumen was used in ancient times as a mortar, and by the Egyptians for embalming.

blast furnace smelting furnace in which the temperature is raised by the injection of an air blast. It is employed in the extraction of metals from their ores, chiefly pig iron from iron ore. The principle has been known for thousands of years, but the present blast furnace is a heavy engineering development combining a number of special techniques.

In the extraction of iron the ingredients of the furnace are iron ore (iron(III) oxide), coke (carbon), and limestone (calcium carbonate).

blast furnace

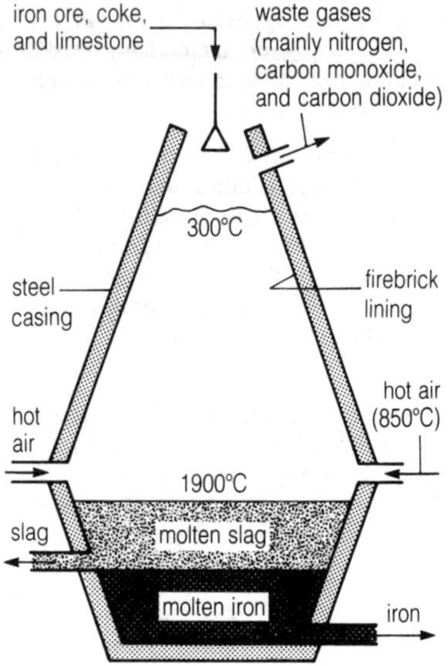

The coke is the fuel and provides the agent for the ◊reduction of the iron ore, carbon monoxide.

$$C + O_2 \rightarrow CO_2$$

$$CO_2 + C \rightarrow 2CO$$

At the high temperature of the furnace iron oxide is reduced to iron.

$$Fe_2CO_3 + 3CO \rightarrow 2Fe + 3CO_2$$

The limestone is decomposed to quicklime (calcium oxide), which combines with the acidic impurities of the ore to form a molten mass known as slag (calcium silicate).

$$CaCO_3 \rightarrow CaO + CO_2$$
$$CaO + SiO_2 \rightarrow CaSiO_3$$

bleaching decolorization of coloured materials. The two main types of bleaching agent are the ***oxidizing bleaches***, which add oxygen and remove hydrogen, and include the ultraviolet rays in sunshine, hydrogen peroxide, and chlorine in household bleaches; and the ***reducing bleaches***, which add hydrogen or remove oxygen, for example sulphur dioxide.

Bleaching processes have been known from antiquity, mainly those acting through sunlight. Both natural and synthetic pigments usually possess highly complex molecules, the colour property often being due only to a part of the molecule. Bleaches usually attack only that small part, yielding another substance similar in chemical structure but colourless.

bleaching powder complex substance made by reacting chlorine with solid slaked lime (calcium hydroxide, $Ca(OH)_2$). It is an oxidizing bleach, the active ingredient of which is the chlorate(I) ion (hypochlorite, OCl^-). Under the action of light this ion slowly releases oxygen.

$$2OCl^- \rightarrow 2Cl^- + O_2$$

This release of oxygen prevents the growth of bacteria. The same reaction makes the addition of chlorine an effective treatment for drinking water and swimming baths, as chlorine and water react to form hypochloric acid (HOCl) and hydrochloric acid.

boiling the process of changing a liquid into its vapour, by heating it at the maximum possible temperature for that liquid (see ◊boiling point) at atmospheric pressure.

boiling point for any given liquid, the temperature at which the application of heat raises the temperature of the liquid no further, but converts it to vapour. The boiling point of water under normal pressure is 100°C/212°F. The lower the pressure, the lower the boiling point and vice versa.

bond the result of the forces of attraction that hold together atoms of an element or elements to form a molecule. The principle types of bonding are ◊ionic, ◊covalent, ◊metallic, and ◊intermolecular (such as hydrogen bonding). The type of bond formed depends on the elements concerned and their electronic structure.

In an ***ionic*** or ***electrovalent bond***, common among inorganic compounds, the combining atoms gain or lose electrons to become ions; for example, sodium (Na) loses an electron to form a sodium ion (Na^+) while chlorine (Cl) gains an electron to form a chloride ion (Cl^-) in the ionic bond of sodium chloride (NaCl).

In a ***covalent bond***, the atomic orbitals of two atoms overlap to form a molecular orbital containing two electrons, which are thus effectively shared between the two atoms. Covalent bonds are common in organic compounds, such as the four carbon–hydrogen bonds in methane (CH_4).

In a ***dative covalent*** or ***coordinate bond***, one of the combining atoms supplies both of the valence electrons in the bond.

In a ***hydrogen bond***, a hydrogen atom joined to an electronegative atom such as nitrogen or oxygen becomes partially positively charged, and is weakly attracted to another electronegative atom on a neighbouring molecule.

The ***metallic bond*** joins metals in a crystal lattice; the atoms occupy lattice positions as positive ions, and valence electrons are shared between all the ions in an 'electron gas'.

boric acid or ***boracic acid*** H_3BO_3 acid formed by the combination of hydrogen and oxygen with non-metallic boron. It is a weak antiseptic and is used in the manufacture of glass and enamels. It is also an efficient insecticide against cockroaches.

boron non-metallic element, symbol B, atomic number 5, relative atomic mass 10.811. In nature it is found only in compounds, as with sodium and oxygen in borax. It exists in two allotropic forms: a brown amorphous powder and very hard, brilliant crystals. Its compounds are used in the preparation of boric acid, water softeners, soaps, enamels, glass, and pottery glazes. In alloys it is used to harden steel. Because it absorbs slow neutrons, it is used to make boron carbide control rods for nuclear reactors. It is a necessary trace element in the human diet.

Boyle's law law stating that the volume of a given mass of gas at a constant temperature is inversely proportional to its pressure. For example, if the pressure of a gas doubles, its volume will be reduced by a half, and vice versa. The law was discovered in 1662 by Robert Boyle.

brass metal ◊alloy of copper and zinc, with not more than 5–6% of other metals. The zinc content ranges from 20% to 45%, and the colour of brass varies accordingly from coppery to whitish yellow. Brasses are characterized by the ease with which they may be shaped and machined; they are strong and ductile, resist many forms of corrosion, and are used for electrical fittings, ammunition cases, screws, household fittings, and ornaments.

Breathalyzer instrument for on-the-spot checking by the police of the amount of alcohol consumed by a suspect driver, who breathes into a plastic bag connected to a tube containing a chemical that changes colour; the chemical used is dilute potassium dichromate in 50% sulphuric acid, which changes from orange to green in the presence of alcohol. Another method is to use a gas chromatograph, again from a breath sample.

brewing the making of beer, ale, or other alcoholic beverage from malt and barley by steeping (mashing), boiling, and fermenting. The term is also used to describe the industry that makes all alcoholic drinks.

Mashing the barley releases the sugar maltose. Yeast is then added, which contains the enzymes needed to convert the maltose into ethanol (alcohol) and carbon dioxide. Hops are added to give a bitter taste.

brine common name for a solution of sodium chloride (NaCl) in water. Brines are used extensively in the food manufacturing industry for canning vegetables, pickling vegetables (sauerkraut manufacture), and curing meat. Industrially, brine is the source from which chlorine, caustic soda (sodium hydroxide), and sodium carbonate are made.

bromide salt of the halide series containing the Br^- ion, which is formed when a bromine atom gains an electron.

When a silver nitrate solution is added to any bromide solution containing dilute nitric acid, a pale yellow precipitate of silver bromide is formed.

$$NaBr_{(aq)} + AgNO_{3\,(aq)} \rightarrow AgBr_{(s)} + NaNO_{3\,(aq)}$$

When chlorine is passed into a solution of a bromide salt, the colourless solution turns red-brown as bromine is produced in a ◊displacement reaction.

$$Cl_{2\,(g)} + 2Br^-_{(aq)} \rightarrow 2Cl^-_{(aq)} + Br_{2\,(l)}$$

The term 'bromide' is sometimes used to describe an organic compound containing a bromine atom, even though it is covalently bonded. Modern naming uses the term 'bromo-' to indicate the covalent nature. For example, the compound C_2H_5Br is now called bromoethane; its traditional name, sometimes still used, is ethyl bromide.

bromine (Greek *bromos* 'stench') dark reddish-brown, non-metallic element, a volatile liquid at room temperature, symbol Br, atomic number 35, relative atomic mass 79.904. It is a member of the ◊halogen group, has an unpleasant odour, and is very irritating to mucous membranes. Its salts are known as bromides.

Bromine was formerly extracted from salt beds, but is now mostly obtained from sea water, where it occurs in small quantities. Its compounds are used in photography and in the chemical and pharmaceutical industries.

bromine water red-brown solution of bromine in water. It is used to test for unsaturation in organic compounds, as such compounds decolourize the solution as they react with the bromine.

Brownian motion continuous random motion of particles in a fluid medium (gas or liquid) as they are subjected to impact from the molecules of the medium. The phenomenon was explained by Albert Einstein in 1905 but was observed as long ago as 1827 by the Scottish botanist Robert Brown (1773–1858). It provides evidence for the ◊kinetic theory of matter.

brown ring test analytical chemistry test for the detection of ◊nitrates. To an aqueous solution containing the test substance is added iron(II) sulphate. Concentrated sulphuric acid is carefully added down the inside wall of the test tube so that it forms a distinct layer at the bottom. The formation of a brown colour at the boundary between the two layers indicates the presence of nitrate.

buffer mixture of chemical compounds chosen to maintain a steady ◊pH.

The commonest buffers consist of a mixture of a weak organic acid and one of its salts or a mixture of ◊acid salts of phosphoric acid. The addition of either an acid or a base causes a shift in the ◊chemical equilibrium, thus keeping the pH constant.

Bunsen Robert Wilhelm von 1811–1899. German chemist, credited with the invention of the ***Bunsen burner***. His name is also given to the carbon–zinc electric cell, which he invented in 1841 for use in arc lamps. In 1859 he discovered two new elements, caesium and rubidium.

burette apparatus used in ◊titrations for the controlled delivery of measured variable quantities of a liquid. It consists of a long, narrow, calibrated glass tube, with a tap at the bottom, leading to a narrow-bore exit.

burning common name for ◊combustion.

butadiene or ***buta-1,3-diene*** $CH_2{=}CHCH{=}CH_2$ inflammable gas derived from petroleum, used in making synthetic rubber and resins.

butane C_4H_{10} one of two gaseous alkanes (paraffin hydrocarbons) having the same formula but differing in structure. Normal butane ($CH_3CH_2CH_2CH_3$) is derived from natural gas; isobutane (2-methylpropane, $CH_3CH(CH_3CH_3)$) is a by-product of petroleum manufacture. Liquefied under pressure, it is used as a fuel for industrial and domestic purposes (for example in portable cookers).

butene C_4H_8 fourth member of the ◊alkene series of hydrocarbons. It is an unsaturated compound, containing one double bond.

by-product substance formed incidentally during the manufacture of some other substance; for example, slag is a by-product of the production of iron in the ◊blast furnace. For industrial processes to be economical, by-products must be recycled or used in other ways as far as possible; in this example, slag is used for making roads.

C

cadmium soft, silver-white, ductile and malleable, metallic element, symbol Cd, atomic number 48, relative atomic mass 112.40. Cadmium occurs in nature as a sulphide or carbonate in zinc ores. It is a toxic metal that, because of industrial dumping, has become an environmental pollutant. Its uses include batteries, electroplating, and as a constituent of alloys used for bearings with low coefficients of friction; it is also a constituent of an alloy with a very low melting point.

Cadmium is also used in control rods for nuclear reactors, owing to its high absorption of neutrons. It was named in 1817 by German chemist Friedrich Strohmeyer (1776–1835) for Cadmus, in Greek mythology ruler of Thebes, near where ores containing it were found.

caesium (Latin *caesius* 'bluish-grey') soft, silvery-white, ductile, metallic element, symbol Cs, atomic number 55, relative atomic mass 132.905. It is one of the ◊alkali metals, and is the most electropositive of all the elements. In air it ignites spontaneously, and it reacts vigorously with water. It is used in the manufacture of photoelectric cells. The name comes from the blueness of its spectral line.

The rate of vibration of caesium atoms is used as the standard of measuring time. Its radioactive isotope Cs-137 (half-life 30.17 years) is one of the most dangerous waste products of the nuclear industry; it is a highly radioactive biological analogue for potassium, produced as a fission product of nuclear explosions and in the reactors of nuclear power plants.

caffeine $C_8H_{10}N_4O_2$ bitter, crystalline substance found in tea, coffee, and kola nuts, that stimulates the heart and central nervous system. Too much caffeine (more than six average cups of coffee a day) can be detrimental to health.

calcination the ◊oxidation of metals by burning in air.

calcium (Latin *calcis* 'lime') soft, silvery-white, metallic element, symbol Ca, atomic number 20, relative atomic mass 40.08. It is one of the ◊alkaline-earth metals. One of the most widely distributed elements, it is the fifth most abundant element (the third most abundant metal) in the Earth's crust. It is found mainly as its carbonate $CaCO_3$, which occurs in a fairly pure condition as chalk and limestone (see ◊calcite). Calcium is an essential component of bones, teeth, shells, milk and leaves, and it forms 1.5% of the human body by mass. Calcium ions in animal cells are involved in regulating muscle contraction, hormone secretion, digestion, and glycogen metabolism in the liver.

The element was discovered by English chemist Humphry Davy in 1808. Its compounds include slaked lime (calcium hydroxide, $Ca(OH)_2$); plaster of Paris (calcium sulphate, $CaSO_4.2H_2O$); calcium hypochlorite ($CaOC1_2$), a bleaching agent; calcium nitrate ($Ca(NO_3)_2.4H_2O$), a nitrogenous fertilizer; calcium carbide (CaC_2), which reacts with water to give ethyne; calcium cyanamide ($CaCN_2$), the basis of many pharmaceuticals, fertilizers, and plastics, including melamine; calcium cyanide ($Ca(CN)_2$), used in the extraction of gold and silver and in electroplating; and others used in baking powders and fillers for paints.

calcium carbonate $CaCO_3$ white solid, found in nature as limestone, marble, and chalk. In its reactions it is a typical ◊carbonate. It is a valuable resource, used in the making of iron, steel, cement, glass, slaked lime, bleaching powder, sodium carbonate and bicarbonate, and many other industrially useful substances. However, its quarrying causes scarring of the landscape, creating dust, noise, and additional traffic in what are often the most attractive areas of the countryside.

calcium hydrogencarbonate $Ca(HCO_3)_2$ substance found in ◊hard water, formed when rainwater passes over limestone rock.

$$CaCO_{3\,(s)} + CO_{2\,(g)} + H_2O_{(l)} \rightarrow Ca(HCO_3)_{2\,(aq)}$$

When this water is boiled it reforms calcium carbonate, removing the hardness; this type of hardness is therefore known as temporary hardness.

calcium hydrogenphosphate $Ca(H_2PO_4)_2$ substance made by heating calcium phosphate with 70% sulphuric acid. It is more soluble in water than calcium phosphate, and is used as a fertilizer.

calcium hydroxide $Ca(OH)_2$ or ***slaked lime*** white solid, slightly soluble in water. A solution of calcium hydroxide is called ◊limewater and is used in the laboratory to test for the presence of carbon dioxide. It is manufactured industrially by adding water to calcium oxide (quicklime) in a strongly exothermic reaction.

$$CaO + H_2O \rightarrow Ca(OH)_2$$

It is used to reduce soil acidity and as a cheap alkali in many industrial processes.

calcium nitrate $Ca(NO_3)_2$ white, crystalline compound that is very soluble in water. The solid decomposes on heating to form oxygen, brown nitrogen(IV) oxide gas and the white solid calcium oxide.

calcium oxide or ***quicklime*** CaO white solid compound, formed by heating ◊calcium carbonate.

$$CaCO_3 \rightarrow CaO + CO_2$$

When water is added it forms calcium hydroxide (slaked lime) in an exothermic reaction.

$$CaO + H_2O \rightarrow Ca(OH)_2$$

It is a typical basic oxide, turning litmus blue.

calcium phosphate or ***calcium orthophosphate*** $Ca_3(PO_4)_2$ white solid, the main constituent of animal bones. It occurs naturally as the mineral apatite and in rock phosphate. It is used in the preparation of phosphate fertilizers.

calcium sulphate $CaSO_4$ white, solid compound, found in nature as gypsum. It dissolves slightly in water to form ◊hard water; this hardness is not removed by boiling, and hence is sometimes called permanent hardness.

calcium superphosphate common name for ◊calcium hydrogenphosphate.

Cannizzaro Stanislao 1826–1910. Italian chemist who revived interest in the work of Avogadro 1811 that had revealed the difference between ◊atoms and ◊molecules, and so established relative atomic and molecular masses as the basis of chemical calculations.

carbide compound of carbon and one other chemical element, usually a metal, silicon, or boron.

Calcium carbide (CaC_2) can be used as the starting material for many basic organic chemical syntheses, by the addition of water and generation of ethyne (acetylene). Some metallic carbides are used in engineering because of their extreme hardness and strength. Tungsten carbide is an essential ingredient of carbide tools and high-speed tools. The 'carbide process' was used during World War II to make organic chemicals from coal rather than from oil.

carbohydrate chemical compound composed of carbon, hydrogen, and oxygen, with the basic formula $C_m(H_20)_n$, and related compounds with the same basic structure but modified ◊functional groups.

The simplest carbohydrates are sugars (***monosaccharides***, such as glucose and fructose, and ***disaccharides***, such as sucrose), which are soluble compounds, some with a sweet taste. When these basic sugar units are joined together in long chains they form ***polysaccharides***, such as starch and glycogen, which often serve as food stores in living organisms. As such they form a major energy-providing part of the human diet. Even more complex carbohydrates are known, including ◊chitin, which is found in the cell walls of fungi and the hard outer skeletons of insects, and ◊cellulose, which makes up the cell walls of plants. Carbohydrates form the chief foodstuffs of herbivorous animals.

carbon (Latin *carbo, carbonaris* 'coal') non-metallic element, symbol C, atomic number 6, relative atomic mass 12.011. It occurs on its own as diamond and graphite (the allotropes), in carbonaceous rocks such as chalk and limestone, as carbon dioxide in the atmosphere, as hydrocarbons in petroleum, coal, and natural gas, and as a constituent of all organic substances.

In its amorphous form, it is familiar as coal, charcoal, and soot. Of the inorganic carbon compounds, the chief one is ***carbon dioxide***, a

carbon

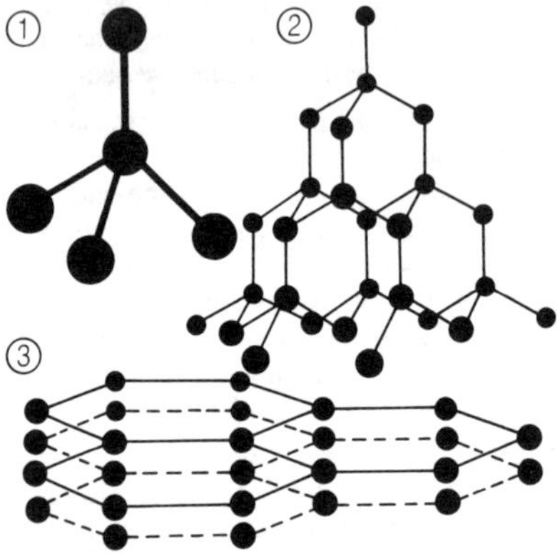

1. the basic unit of the diamond structure
2. *diamond*, a giant three-dimensional structure
3. *graphite*, a two-dimensional structure

colourless gas formed when carbon is burned in an adequate supply of air, and ***carbon monoxide*** (CO), formed when carbon is oxidized in a limited supply of air. ***Carbon disulphide*** (CS_2) is a dense liquid with a sweetish odour. Another group of compounds is the ***carbon halides***, including ◊carbon tetrachloride (CCl_4).

When added to steel, carbon forms a wide range of alloys with useful properties. In pure form, it is used as a moderator in nuclear reactors; as colloidal graphite it is a good lubricant and, when deposited on a surface in a vacuum, obviates photoelectric and secondary emission of electrons. Carbon is used as a fuel in the form of coal or coke. The radioactive isotope carbon-14 (half-life 5,730 years) is used as a tracer in biological research.

The element has the following characteristic reactions:

with air or oxygen It burns on heating to form carbon dioxide in excess air, or carbon monoxide in a limited supply of air.

$$C + O_2 \rightarrow CO_2$$
$$\Delta H = -394 \text{ kJ mol}^{-1}$$

$$2C + O_2 \rightarrow 2CO$$

with metal oxides It reduces many metal oxides at high temperatures.

$$Fe_2O_3 + 3C \rightarrow 2Fe + 3CO$$

with steam It forms water gas (a cheap, useful, industrial fuel) when steam is passed over white-hot coke.

$$C + H_2O \rightarrow CO + H_2$$

with concentrated acids With hot, concentrated sulphuric or nitric acids it forms carbon dioxide.

carbonate CO_3^{2-} ion formed when carbon dioxide dissolves in water, and any salt formed by this ion and another chemical element, usually a metal.

The carbon dioxide (CO_2) dissolved by rain falling through the air, and liberated by decomposing animals and plants in the soil, forms with water carbonic acid (H_2CO_3), which unites with various basic substances to form carbonates. Calcium carbonate ($CaCO_3$) (chalk, limestone, and marble) is one of the most abundant carbonates known, being a constituent of mollusc shells and the hard outer skeletons of crustaceans.

Carbonates give off carbon dioxide when heated or treated with dilute acids. This is used as the laboratory test for the presence of the ion, as it gives an immediate effervescence, with the gas turning lime-water (a solution of calcium hydroxide, $Ca(OH)_2$) milky. See ◊sodium carbonate and ◊calcium carbonate.

carbonated water water in which carbon dioxide is dissolved under pressure. It forms the the basis of many fizzy soft drinks such as soda water and lemonade.

carbon cycle the sequence by which ◊carbon circulates and is recycled through the natural world. The carbon element from carbon dioxide, released into the atmosphere by animals as a result of ◊respiration, is taken up by plants during ◊photosynthesis and converted into carbohydrates; the oxygen component is released back into the atmosphere.

carbon cycle

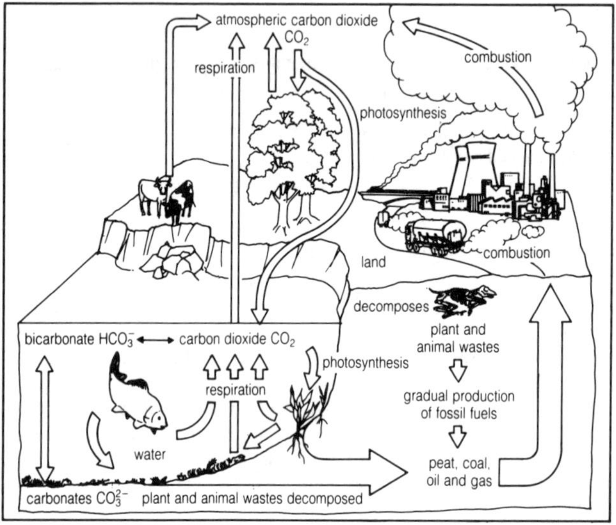

The simplest link in the carbon cycle, however, occurs when an animal eats a plant and carbon is transferred from, say, a leaf cell to the animal body. Today, the carbon cycle is being disrupted by the increased consumption and burning of fossil fuels, and the burning of large tracts of tropical forests, as a result of which levels of carbon dioxide are building up in the atmosphere and probably contributing to the ◊greenhouse effect.

carbon dating alternative name for ◊radiocarbon dating.

carbon dioxide CO_2 colourless gas, slightly soluble in water and denser than air. It is formed when carbon and carbon-containing compounds are fully oxidized, as when they are burnt in an excess of air. It is also produced when acids are added to carbonates or hydrogencarbonates, and when these salts are heated. It is a typical acidic oxide, dissolving in water to give a solution of the weak dibasic acid carbonic acid, and forming salts with alkalis.

$$H_2O + CO_2 \leftrightarrow H_2CO_3 \leftrightarrow H^+_{(aq)} + HCO_3^-{}_{(aq)}$$

$$NaOH + CO_2 \rightarrow NaHCO_3$$

With a solution of calcium hydroxide (limewater) the gas forms a white (milky) precipitate of calcium carbonate. This reaction is used as the confirmatory test for carbon dioxide.

$$Ca(OH)_{2\,(aq)} + CO_{2\,(g)} \rightarrow CaCO_{3\,(s)} + H_2O_{(l)}$$

The gas supports the combustion of burning magnesium, but extinguishes lower-temperature flames; it is used in fire extinguishers.

Carbon dioxide plays a vital role in the carbon cycle, in the formation of ◊hard water, and in the ◊greenhouse effect. It is used as a coolant in its solid form (known as 'dry ice'), and in the chemical industry.

carbonic acid H_2CO_3 weak, dibasic acid formed by dissolving carbon dioxide in water.

$$H_2O + CO_2 \leftrightarrow H_2CO_3$$

It forms two series of salts: ◊carbonates and ◊hydrogencarbonates. Fizzy drinks are made by dissolving carbon dioxide in water under pressure; soda water is a solution of carbonic acid.

carbon monoxide CO colourless, odourless gas formed when carbon is oxidized in a limited supply of air. It is a poisonous constituent of car exhaust fumes, forming a stable compound with haemoglobin in the blood, thus preventing the haemoglobin from transporting oxygen to the body tissues.

In industry it is used as a reducing agent in metallurgical processes, such as the extraction of iron in a ◊blast furnace. It is a constituent of cheap industrial fuels such as water gas (see ◊carbon). In air it burns with a luminous blue flame to form carbon dioxide.

$$2CO + O_2 \rightarrow 2CO_2$$

carbon tetrachloride common name for ◊tetrachloromethane.

Carborundum trademark for a very hard, black abrasive, consisting of silicon carbide (SiC), an artificial compound of carbon and silicon. First produced in 1891, it is harder than ◊corundum but not as hard as ◊diamond.

carboxylic acid R-COOH organic acid containing the carboxyl group (COOH) attached to another group (R), which can be hydrogen (giving methanoic acid, HCOOH) or a larger molecule (up to 24 carbon atoms). The smaller carboxylic acids form a homologous series, with all the names ending -oic (methanoic acid, HCOOH; ethanoic acid, CH_3COOH; propanoic acid, C_2H_5COOH, and so on). Larger ones are often found as ◊esters in fats, often with glycerine, and so are called ◊fatty acids.

cast iron cheap but invaluable constructional material, most commonly used for car engine blocks. Cast iron is partly refined pig (crude) ◊iron, which is very fluid when molten and highly suitable for shaping by casting, as it contains too many impurities, such as carbon, to be readily shaped in any other way. Solid cast iron is heavy and can absorb great shock but is very brittle.

catalyst substance that alters the speed of or makes possible a chemical or biochemical reaction but that remains unchanged at the end of the reaction. ◊Enzymes are natural biochemical catalysts. In practice most catalysts are used to speed up reactions.

catalytic converter device for reducing toxic emissions from the internal-combustion engine. It converts harmful exhaust products to relatively harmless ones by passing exhaust gases over a mixture of catalysts. ***Oxidation catalysts*** convert hydrocarbons into carbon dioxide and water; ***three-way catalysts*** convert oxides of nitrogen back into nitrogen. Catalytic converters are standard in the USA, where a 90% reduction in pollution from cars was achieved without loss of engine performance or fuel economy.

catalytic cracking a form of ◊cracking.

cathode the negative electrode towards which positive particles (cations) move within a device such as a cell in a battery, an electrolytic cell, or a diode.

cation ◊ion carrying a positive charge. During electrolysis, cations in the electrolyte move to the cathode (negative electrode).

caustic soda former name for ◊sodium hydroxide.

cell, chemical apparatus that produces electrical energy as a result of a chemical reaction; the popular name is 'battery', but this actually refers to a collection of cells in one unit. A ***primary*** electric cell cannot be replenished, whereas in a ***secondary*** cell or accumulator, the action is reversible and the original condition can be restored by an electric current.

The reactions that take place in a simple cell depend on the fact that some metals are more reactive than others. If two different metals are joined by an ◊electrolyte and a wire, the more reactive metal loses electrons to form ions. The ions pass into solution in the electrolyte, while the electrons flow down the wire to the less reactive metal. At the less reactive metal the electrons are taken up by the positive ions in the electrolyte, which completes the circuit. If the two metals are zinc and copper and the electrolyte is dilute sulphuric acid, the following cell reactions occur.

The zinc atoms dissolve as they lose electrons (oxidation).

$$Zn - 2e^- \rightarrow Zn^{2+}$$

The two electrons travel down the wire and are taken up by the hydrogen ions in the electrolyte (reduction).

$$2H^+ + 2e^- \rightarrow H_2$$

The overall cell reaction is obtained by combining these two reactions; the zinc rod slowly dissolves and bubbles of hydrogen appear at the copper rod.

$$Zn + 2H^+ \rightarrow Zn^{2+} + H_2$$

If each rod is immersed in an electrolyte containing ions of that metal, and the two electrolytes are joined by a salt bridge, metallic copper deposits on the copper rod as the zinc rod dissolves in a ◊redox reaction, just as if zinc had been added to a copper salt solution.

$$Zn - 2e^- \rightarrow Zn^{2+}$$

cell, chemical

the basic principle of a chemical cell

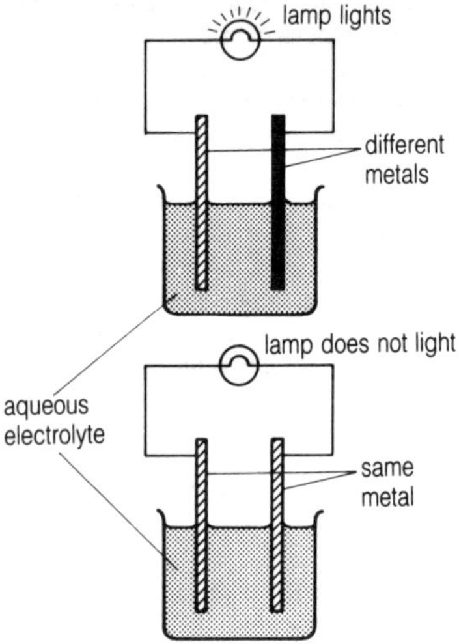

a simple cell

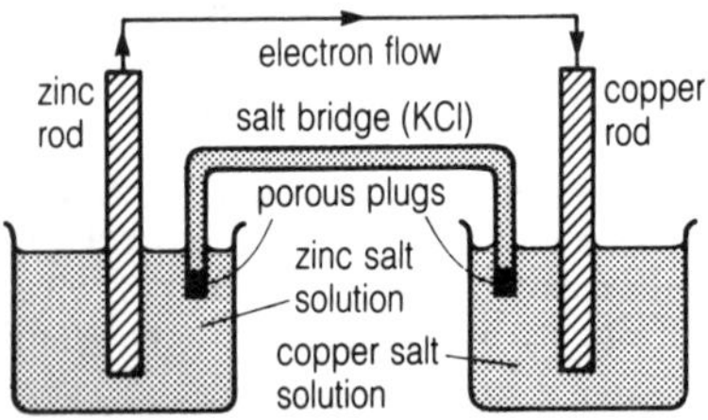

$$Cu^{2+} + 2e^- \rightarrow Cu$$

$$Zn_{(s)} + Cu^{2+}_{(aq)} \rightarrow Zn^{2+}_{(aq)} + Cu_{(s)}$$

This mechanism is the basis of the sacrificial protection of metals (see ◊rust prevention).

celluloid transparent or translucent, highly inflammable, plastic material (a ◊thermoplastic made from nitrocellulose and camphor) once used for toilet articles, novelties, and photographic film. It has been replaced by the non-inflammable substance cellulose acetate.

Celsius scale temperature scale in which one division or degree is taken as one hundredth part of the interval between the freezing point (0°C) and the boiling point (100°C) of water at standard atmospheric pressure. The degree centigrade (°C) was officially renamed Celsius in 1948 to avoid confusion with the angular measure known as the centigrade (one hundredth of a grade). The Celsius scale is named after the Swedish astronomer Anders Celsius (1701–44), who devised it in 1742, but in reverse (freezing point was 100°; boiling point 0°).

cement any bonding agent used to unite particles in a single mass or to cause one surface to adhere to another. ***Portland cement*** is a powder obtained from burning together a mixture of lime (or chalk) and clay; mixed with water and sand or gravel, it turns into mortar or concrete.

centigrade scale common name for the ◊Celsius temperature scale.

centrifuge apparatus that rotates containers at high speeds, creating centrifugal forces. One use is for separating mixtures of substances of different densities.

The mixtures are placed in the containers and the rotation sets up centrifugal forces, causing them to separate according to their densities. A common example is the separation of the lighter plasma from the heavier blood corpuscles in certain blood tests. The ***ultracentrifuge*** is a very high-speed centrifuge, used in biochemistry for separating ◊colloids and organic substances; it may operate at several million revolutions per minute. Large centrifuges are used for physiological research – for example, in astronaut training where bodily response to many times the normal gravitational force is tested.

CFC abbreviation for ◊chlorofluorocarbon.

chain reaction mechanism that produces very fast, ◊exothermic reactions, as in the formation of flames and explosions.

The reaction begins with the formation of a single reactive molecule. This combines with an inactive molecule to form two reactive molecules. These two produce four (or more) reactive molecules; very quickly, very many reactive molecules are produced, so the reaction rate accelerates dramatically. The reactive molecules contain an unpaired electron and are called ◊free radicals; they last only a short time because they are so reactive.

In a ***nuclear chain reaction***, the fission of one nucleus results in the emission of particles that collide with other nuclei and cause them to split in turn.

chalk soft, fine-grained, whitish rock composed of calcium carbonate $CaCO_3$, extensively quarried for use in cement, lime, and mortar, and in the manufacture of cosmetics and toothpaste. ***Blackboard chalk*** in fact consists of ◊gypsum (calcium sulphate, $CaSO_4$).

change of state the change that takes place when a gas condenses to a liquid, a liquid freezes to a solid, a solid melts to form a liquid or a liquid vaporizes (evaporates) to form a gas. The first set of changes are brought about by cooling, the second set by heating. In the unusual change of state called ***sublimation***, a solid changes directly to a gas without passing through the liquid state. For example, solid carbon dioxide (dry ice) sublimes to carbon dioxide gas.

charcoal black, porous form of ◊carbon, produced by heating wood or other organic materials in the absence of air (a process called destructive distillation). It is used as a fuel, for smelting metals such as copper and zinc; in the form of ***activated charcoal***, for purifying and filtration of drinking water and other liquids and gases; and by artists for making black line drawings.

Charcoal was traditionally produced by burning dried wood in a kiln, a process lasting several days. The kiln was either a simple hole in the ground, or an earth-covered mound. Today kilns are of brick or iron, both of which allow the waste gases to be collected and used.

change of state

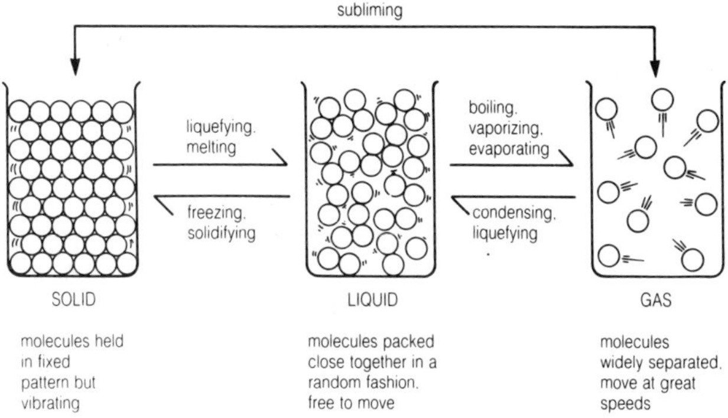

Charcoal had many uses in earlier centuries. Because of the high temperature at which it burns (1,100°C), it was used in furnaces and blast furnaces before the development of ◊coke. It was also used in an industrial process for obtaining ethanoic acid, in producing wood tar and wood pitch, and (when produced from alder or willow trees) as a component of gunpowder.

Charles's law law stating that the volume of a given mass of gas at constant pressure is directly proportional to its absolute temperature (temperature in kelvin). It was discovered by Jacques Charles in 1787 and independently by Joseph Gay-Lussac in 1802.

chemical change the change that occurs in a chemical reaction. A new substance is formed, with properties different from those of the starting materials; significant energy changes are involved; and the change cannot easily be reversed. A simple example of chemical change is the burning of carbon.

chemical element alternative name for ◊element.

chemical equation method of indicating the reactants and products of a chemical reaction by using chemical symbols and formulae. A chemical equation gives two basic pieces of information: (1) the reactants (on the left-hand side) and products (right-hand side); and (2) the reacting proportions (stoichiometry), that is how many units of each reactant and product are involved. The equation must balance; that is, the total number of atoms of a particular element on the left-hand side must be the same as the number of atoms of that element on the right-hand side.

$$\underset{\text{reactants}}{Na_2CO_3 + 2HCl} \rightarrow \underset{\text{products}}{2NaCl + CO_2 + H_2O}$$

This equation states that one molecule of sodium carbonate combines with two molecules of hydrochloric acid to from two molecules of sodium chloride, one of carbon dioxide, and one of water. ◊State symbols and the energy symbol (ΔH) can be used to provide further information.

$$Na_2CO_{3\,(s)} + 2HCl_{(aq)} \rightarrow 2NaCl_{(aq)} + CO_{2\,(g)} + H_2O_{(l)}\ (-\Delta H)$$

Substituting the relative molecular masses of the substances indicates the proportions of masses involved.

Double arrows indicate that the reaction is reversible – in the formation of ammonia from hydrogen and nitrogen, the direction depends on the temperature and pressure of the reactants.

$$3H_2 + N_2 \leftrightarrow 2NH_3$$

chemical equilibrium condition in which the products of a reversible chemical reaction are formed at the same rate at which they decompose back into the reactants, so that the concentration of each reactant and product remains constant.

chemical family collection of elements that have very similar chemical and physical properties. In the ◊periodic table of the elements such collections are to be found in the vertical columns (groups). The groups that contain the most markedly similar elements are group I, the ◊alkali metals; group II, the ◊alkaline-earth elements; group VII, the ◊halogens; and group 0, the noble or ◊inert gases.

chemisorption the attachment, by chemical means, of a single layer of molecules, atoms, or ions of gas to the surface of a solid or, less frequently, a liquid. It is the basis of catalysis (see ◊catalyst) and of great industrial importance.

chemistry the science concerned with the composition of matter and of the changes that take place in it under certain conditions.

Examination and possible breakdown of compounds to determine their components is ***analysis***, and the building up of compounds from their components is ***synthesis***. When substances are brought together without changing their molecular structures they are said to be ***mixtures***. Chemical compounds are produced by a chemical action that alters the arrangement of the atoms in the molecule. Heat, light, vibration, catalytic action, radiation, or pressure, as well as moisture (for ionization), may be necessary to produce a chemical change.

Organic chemistry is the branch of chemistry that deals with carbon compounds. ***Inorganic chemistry*** deals with the description, properties, reactions, and preparation of all the elements and their compounds, with the exception of carbon compounds. ***Physical chemistry*** is concerned with the quantitative explanation of chemical phenomena and reactions, and the measurement of data required for such explanations. This branch studies in particular the movement of molecules and the effects of temperature and pressure, often with regard to gases and liquids.

history Ancient civilizations were familiar with certain chemical processes – for example, extracting metals from their ores, and making alloys. The alchemists endeavoured to turn base metals into gold, and chemistry evolved towards the end of the 17th century from the techniques and insights developed during alchemical experiments. Robert Boyle defined elements as the simplest substances into which matter could be resolved. The alchemical doctrine of the four elements (earth, air, fire, and water) gradually lost its hold, and the theory that all combustible bodies contain a substance called phlogiston (a weightless 'fire element' generated during combustion) was discredited in the 18th century by the experimental work of Black, Lavoisier, and Priestley (who discovered the presence of oxygen in air). Cavendish discovered the composition of water, and Dalton put forward the atomic theory, which ascribed a precise relative weight to the 'simple atom' characteristic of each element. Much research then took place leading to the development of ◊biochemistry, chemotherapy, and ◊plastics.

chlorate any salt from an acid containing both chlorine and oxygen (ClO, ClO_2, ClO_3, and ClO_4). Common chlorates are those of sodium, potassium, and barium. Certain chlorates are used in weedkillers.

chloride Cl^- negative ion formed when hydrogen chloride dissolves in water, and any salt containing this ion, commonly formed by the action of hydrochloric acid (HCl) on various metals or by direct combination of a metal and chlorine. Sodium chloride ($NaCl$) is common table salt.

chlorinated solvents liquid organic compounds that contain chlorine atoms, often two or more. They are very effective solvents for fats and greases, but many have toxic properties; they include chloroform ($CHCl_3$), carbon tetrachloride (CCl_4), and trichloroethene ($CH_2ClCHCl_2$).

chlorine greenish-yellow, gaseous, non-metallic element with a pungent odour, symbol Cl, atomic number 17, relative atomic mass 35.453 (a mixture of two isotopes: 75% Cl-35, 25% Cl-37). It is the second member of the ◊halogen group (group VII of the periodic table). In nature it is widely distributed in combination with the ◊alkali metals, as chlorates or chlorides; in its pure form the gas is a diatomic molecule (Cl_2). It is a very reactive element, and combines with most metals, some non-metals, and a wide variety of compounds.

Industrially, chlorine is prepared by electrolysis of brine (see ◊sodium hydroxide). It is made on a large scale and used in making bleaches, for sterilizing water for drinking and in swimming baths, and in the manufacture of chloro-organic compounds such as chlorinated solvents, CFCs, and PVC. Some typical reactions are given below.

with metals When dry chlorine is passed over a heated metal, the chloride is formed.

$$Zn + Cl_2 \rightarrow ZnCl_2$$

with non-metals The same reaction occurs with certain non-metals, when the dry gas is passed over the heated element.

$$2P + 5Cl_2 \rightarrow 2PCl_5$$

with compounds With water, chlorine forms a bleaching solution.

$$H_2O + Cl_2 \rightarrow HCl + HOCl$$

$$2OCl^- \rightarrow 2Cl^- + O_2$$

Iron (II) salts are oxidized to iron (III) salts.

$$2FeCl_2 + Cl_2 \rightarrow 2FeCl_3$$

Organic compounds undergo halogenation.

$$C_2H_6 + Cl_2 \rightarrow C_2H_5Cl + HCl$$

Alkalis form chlorides, chlorates, and water.

$$2NaOH + Cl_2 \rightarrow NaCl + NaOCl + H_2O$$

Other halogens are displaced in a redox reaction.

$$2KBr + Cl_2 \rightarrow 2KCl + Br_2$$

chlorofluorocarbon (CFC) synthetic chemical, which is odourless, non-toxic, non-flammable, and chemically inert. CFCs are used as propellants in aerosol cans, refrigerants in refrigerators and air conditioners, in the manufacture of foam boxes for take-away food cartons, and as cleaning substances in the electronics industry. They are partly responsible for the destruction of the ◊ozone layer. In June 1990 representatives of 93 nations, including the UK and the USA, agreed to phase out production of CFCs and other ozone-depleting chemicals by the end of the 20th century.

When CFCs are released into the atmosphere, they drift up slowly into the stratosphere, where, under the influence of ultraviolet radiation from the Sun, they break down into chlorine atoms which destroy the ozone layer and allow harmful radiation from the Sun to reach the Earth's surface. CFCs can remain in the atmosphere for more than 100 years. Replacements for CFCs are being developed, and research into safe methods of destroying existing CFCs is being carried out.

chloroform or ***trichloromethane*** $CHCl_3$ clear, colourless, toxic, carcinogenic liquid with a characteristic, pungent, sickly-sweet smell and taste, formerly used as an anaesthetic (now superseded by less harmful substances). It is used as a solvent and in the synthesis of organic chemical compounds.

chlorophyll green pigment present in most plants that is responsible for the absorption of light energy during ◊photosynthesis. The pigment absorbs the red and blue-violet parts of sunlight but reflects the green, thus giving plants their most characteristic colour. It is similar in structure to haemoglobin, but has magnesium in the reactive part of the molecule instead of iron.

chromatography technique for separating a mixture of gases, liquids, or dissolved substances into its constituent components. This is done by passing the mixture (the 'mobile phase') through another substance (the 'stationary phase'), usually a liquid or solid. The different components of the mixture are absorbed or impeded by the stationary phase to different extents, and therefore become separate. The technique is used for both qualitative and quantitive analyses in biology and chemistry.

In ***paper chromatography***, the mixture separates because the components have differing solubilities in the solvent flowing through the paper and in the chemically bound water of the paper.

In ***thin-layer chromatography***, a wafer-thin layer of absorbent medium on a glass plate replaces the filter paper. The mixture separates because of the differing solubilities of the components in the solvent flowing up the solid layer, and their differing tendencies to stick to the solid (adsorbtion). The same principles apply in ***column chromatography***.

In ***gas–liquid chromatography***, a gaseous mixture is passed into a long, coiled tube (enclosed in an oven) filled with an inert powder coated in a liquid. A carrier gas flows through the tube. As the mixture proceeds along the tube it separates as the components dissolve in the liquid to differing extents or stay as a gas. A detector locates the different components as they emerge from the tube. The technique is very powerful, allowing tiny quantities of substances (fractions of parts per million) to be separated and analysed.

Analytical chromatography uses very small quantities, often millionths of a gram or less, to identify and quantify components of a mixture. Examples are the determination of the identities and amounts of amino acids in a protein, and the determination of the alcohol content of blood and urine samples.

chromatography

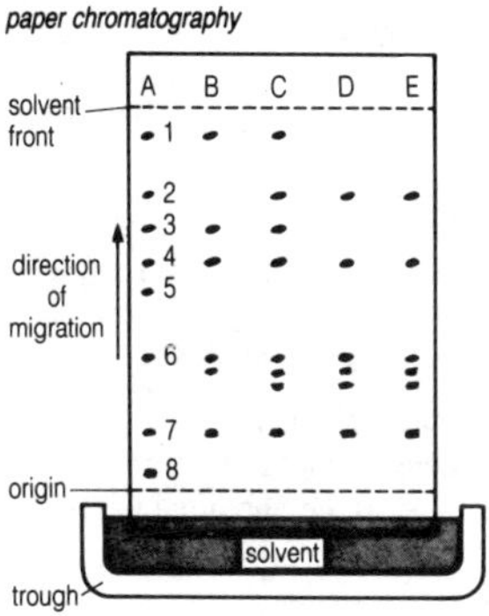

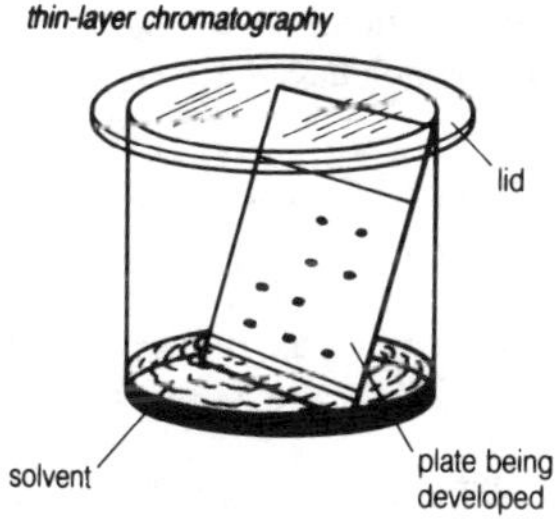

Preparative chromatography is used on a large scale for the purification and collection of one or more of the constituents, for example the recovery of protein from abattoir and other effluent wastes. The technique was first used in the separation of coloured mixtures into their component pigments.

chromium (Greek *chromos* 'colour') hard, brittle, grey-white, metallic element, symbol Cr, atomic number 24, relative atomic mass 51.996.

It takes a high polish, has a high melting point, and is very resistant to corrosion. It is used in chromium electroplating, to make stainless steel and other alloys, and as a catalyst. Its compounds are used for tanning leather and for ◊alums. In human nutrition it is a vital trace element. In nature, it occurs chiefly as a chrome–iron ore ($FeCr_2O_4$).

citric acid $HOOCCH_2C(OH)(COOH)CH_2COOH$ organic acid widely distributed in the plant kingdom, found in high concentrations in citrus fruits, with a sharp, sour taste. At one time it was commercially prepared from concentrated lemon juice, but now the main source is the fermentation of sugar with certain moulds.

coal black or blackish mineral substance of fossil origin, the result of the transformation of ancient plant matter under progressive compression. It is used as a fuel and in the chemical industry.

coal tar black oily material resulting from the destructive distillation of bituminous coal. Further distillation of coal tar yields a number of fractions: light oil, middle oil, heavy oil, and anthracene oil; the residue is called pitch. On further fractionation a large number of substances are obtained, about 200 of which have been isolated. They are used as dyes and in medicines.

cobalt (German *Kobalt* 'goblin') hard, lustrous, grey, metallic element, symbol Co, atomic number 27, relative atomic mass 58.933. It is found in various ores and occasionally as a free metal, sometimes in metallic meteorite fragments. It is used in the preparation of magnetic, wear-resistant, and high-strength alloys; its compounds are used in inks, paints, and varnishes.

The isotope Co-60 is radioactive (half-life 5.3 years) and is produced in large amounts for gamma rays to be used in industrial radiography, research, and cancer treatments. It was named in 1730 by Swedish chemist Georg Brandt (1694–1768); the name derives from the fact that miners considered its ore worthless because of its arsenic content.

cobalt-60 radioactive (half-life 5.3 years) isotope produced by neutron radiation of cobalt in heavy-water reactors, used in large amounts for gamma rays in cancer therapy, industrial radiography, and research, substituting for the much more costly radium.

cobalt chloride $CoCl_2$ compound that exists in two forms: the hydrated salt ($CoCl_2.6H_2O$), which is pink, and the anhydrous salt, which is blue. The anhydrous form is used as a chemical test for water, as it forms the hydrated salt and turns pink when water is present. When the hydrated salt is gently heated the anhydrous salt is reformed.

$$CoCl_2 + 6H_2O \leftrightarrow CoCl_2.6H_2O$$

coke clean, light fuel produced by the carbonization of certain types of coal. When this coal is strongly heated in airtight ovens (in order to release all volatile constituents), the brittle, silver-grey remains are coke. Coke comprises 90% carbon together with very small quantities of water, hydrogen, and oxygen, and makes a useful industrial and domestic fuel.

collision theory theory that explains chemical reactions and the way in which the rate of reaction alters when the conditions alter. For a reaction to occur the reactant particles must collide. Only a certain fraction of the total collisions cause chemical change; these are called ***fruitful collisions***. These fruitful collisions have sufficient energy (activation energy) at the moment of impact to break the existing bonds and form

collision theory

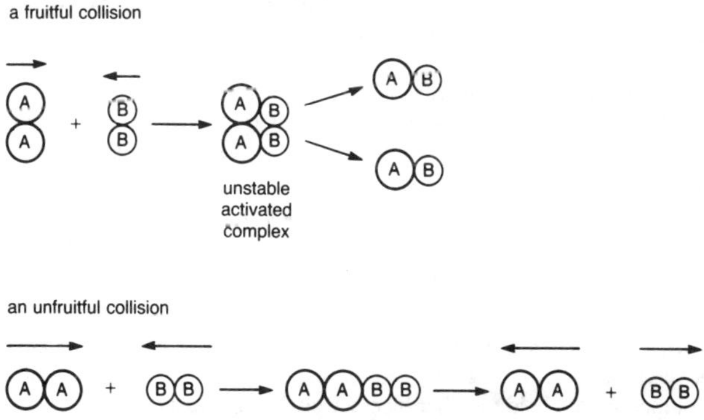

new bonds, resulting in the products of the reaction. Increasing the concentration of the reactants and raising the temperature bring about more collisions and therefore more fruitful collisions, increasing the rate of reaction.

When a ◊catalyst undergoes collision with the reactant molecules, less energy is required at the moment of impact for the chemical change to occur, and hence more collisions have sufficient energy for reaction to occur. The reaction rate therefore increases.

colloid substance composed of extremely small particles whose size is between those in suspension and those in true solution (between 1 and 1,000 microns across). The two components together are the ***continuous phase***; the ***dispersed phase*** is distributed in the former. There are various types of colloids: those involving gases include an aerosol (a dispersion of a liquid or solid in a gas, as in fog or smoke) and a foam (a dispersion of a gas in a liquid). Liquids form both the dispersed and continuous phases in an ***emulsion***.

Milk is a natural emulsion (stable colloidal suspension) of liquid fat in a watery liquid; synthetic emulsions such as some paints and cosmetic lotions have chemical emulsifying agents to stabilize the colloid and stop the two phases from separating out. Colloidal solutions (a solid dispersed in a liquid) are called ***sols***. A sol in which both phases contribute to the molecular three-dimensional network of the colloid take on a jellylike form and are known as ***gels***; gelatine, starch 'solution' and silica gel are common examples. Colloids were first studied thoroughly by British chemist Thomas Graham, who defined them as substances that (in solution) will not diffuse through a semipermeable membrane (as opposed to crystalloids, solutions of inorganic salts, which will diffuse through).

combustion burning, defined in chemical terms as rapid combination of a substance with oxygen accompanied by the evolution of heat and usually light. A slow-burning candle flame and the explosion of a mixture of petrol vapour and air are extreme examples of combustion.

compound chemical substance made up of two or more ◊elements bonded together, so that they cannot be separated by physical means. Compounds are held together by electrovalent or covalent bonds.

concentration the amount of a substance (◊solute) present in a specified amount of a solution. Either amount may be specified as a mass or a volume (liquids only). Common units used are ◊moles per cubic decimetre, grams per cubic decimetre, grams per 100 cubic centimetres, or grams per 100 grams.

The term also refers to the process of increasing the concentration of a solution by removal of some of the substance in which the solute is dissolved (◊solvent). In a ***concentrated*** solution, the solute is present in large quantities. Concentrated brine is around 30% sodium chloride in water; concentrated caustic soda (caustic liquor) is around 40% sodium hydroxide; and concentrated sulphuric acid is 98% acid.

condensation the conversion of a vapour to a liquid as it loses heat. This is frequently achieved by letting the vapour come into contact with a cold surface. It is an essential step in ◊distillation processes.

condensation reaction or ***addition–elimination reaction*** reaction in which two organic compounds combine to form a larger molecule, accompanied by the removal of a smaller molecule (usually water).

condensation polymerization ◊polymerization reaction in which one or more monomers, with more than one reactive functional group, combine to form a polymer with the elimination of water or another small molecule. Polyamides (such as nylon) and polyesters (such as Terylene) are made by condensation polymerization.

condenser laboratory apparatus used to condense vapours back to liquid so that the liquid can be recovered. It is used in ◊distillation and in reactions where the liquid mixture can be kept boiling without the loss of solvent.

conductor material that conducts heat or electricity (as opposed to a non-conductor or insulator). A good conductor has a high electrical or heat conductivity, and is generally a substance rich in free electrons such as a metal. A poor conductor (such as the non-metals glass and porcelain) has few free electrons. Carbon is exceptional in being non-metallic and yet (in some of its forms) a relatively good conductor of heat and electricity. Substances such as silicon and germanium, with intermediate conductivities that are improved by heat, light, or voltage, are known as ◊semiconductors.

conservation of energy principle stating that in a chemical reaction the total amount of energy in the system remains unchanged.

For each component there may be changes in energy due to change of physical state, changes in the nature of chemical bonds, and either an input or output of energy. However, there is no net gain or loss of energy.

conservation of mass principle stating that in a chemical reaction the sum of all the masses of the substances involved in the reaction (reactants) is equal to the sum of all of the masses of the substances produced by the reaction (products) – that is, no matter is gained or lost.

constantan high-resistance alloy of approximately 40% nickel and 60% copper with a very low temperature coefficient. It is used in electrical resistors.

constant composition, law of law stating that the proportions of the amounts of the elements in a pure compound are always the same and are independent of the method used to produce it. This is the basis upon which it is possible to assign a single chemical formula to a pure compound.

contact process the main industrial method of manufacturing the chemical ◊sulphuric acid. Sulphur dioxide and air are passed over a hot (450°C) catalyst of vanadium (V) oxide. The sulphur trioxide produced is then absorbed in concentrated sulphuric acid to make fuming sulphuric acid (oleum). Unreacted gases are recycled. The oleum is diluted with water to give concentrated sulphuric acid (98%).

$$2SO_2 + O_2 \leftrightarrow 2SO_3$$

$$H_2SO_4 + SO_3 \leftrightarrow H_2S_2O_7 \quad \text{(oleum)}$$

$$H_2S_2O_7 + H_2O \rightarrow 2H_2SO_4$$

copper orange-pink, very malleable and ductile, metallic element, symbol Cu (from Latin *cuprum*), atomic number 29, relative atomic mass 63.546. It is used for its toughness, softness, pliability, high thermal and electrical conductivity, and resistance to corrosion.

It was the first metal used systematically for tools by humans; when mined and worked into utensils it was the basis for the Copper Age in prehistory. When alloyed with tin it forms bronze, whch strengthens the copper, allowing it to hold a sharp edge; the systematic production and use of this was the basis for the prehistoric Bronze Age. Brass, another hard copper alloy, includes zinc. The name comes from the Greek for Cyprus (*Kyprios*), where copper was mined.

copper(II) carbonate $CuCO_3$ green solid that readily decomposes to form black copper(II) oxide on heating. It dissolves in dilute acids to give blue solutions with effervescence caused by the giving off of carbon dioxide.

$$CuCO_3 + H_2SO_4 \rightarrow CuSO_4 + CO_2 + H_2O$$

copper(II) oxide CuO black solid that is readily reduced to copper by carbon, carbon monoxide, or hydrogen if heated with any of these.

$$CuO + C \rightarrow Cu + CO$$

$$CuO + CO \rightarrow Cu + CO_2$$

$$CuO + H_2 \rightarrow Cu + H_2O$$

It is usually made in the laboratory by heating copper(II) carbonate, nitrate, or hydroxide.

$$2Cu(NO_3)_2 \rightarrow 2CuO + 4NO_2 + O_2$$

Copper(II) oxide is a typical basic oxide, dissolving readily in most dilute acids.

copper(II) sulphate $CuSO_4$ substance usually found as a blue, crystalline, hydrated salt $CuSO_4.5H_2O$ (also called blue vitriol). It is made from the action of dilute sulphuric acid on copper(II) oxide, hydroxide, or carbonate.

$$CuO + H_2SO_4 + 4H_2O \rightarrow CuSO_4.5H_2O$$

When the hydrated salt is heated gently it loses its water of crystallization and the blue crystals turn to a white powder. The reverse reaction is used as a chemical test for water.

$$CuSO_4.5H_2O \leftrightarrow CuSO_4 + 5H_2O$$

covalency form of ◊valency in which two atoms unite by sharing a pair of electrons (see ◊covalent bond).

covalent bond chemical ◊bond produced when two atoms each contribute an electron for mutual sharing. It is composed of two electrons and is often represented by a single line drawn between the two atoms. This type of bonding always produces a ◊molecule. Covalently bonded substances include hydrogen (H_2) and water (H_2O).

cracking reaction where a large ◊alkane molecule is broken down by heat into a smaller alkane and a small ◊alkene molecule. The reaction is carried out at a high temperature (600°C or higher) and often in the presence of a catalyst.

$$C_{12}H_{26} \rightarrow C_8H_{18} + 2C_2H_4$$

Cracking is a commonly used process in the petrochemical industry. It is the main method of preparation of alkenes and is also used to manufacture petrol from the higher-boiling-point fractions obtained from the ◊fractionation of crude oil.

creosote black, oily liquid derived from coal tar, used as a wood preservative. Medicinal creosote, which is transparent and oily, is derived from wood tar.

cross linking sideways linking between two or more long-chain molecules in a ◊polymer. Cross linking gives the polymer a higher melting point and makes it harder; Examples of cross-linked polymers include Bakelite and vulcanized rubber.

crude oil the unrefined form of ◊petroleum.

cryolite rare granular crystalline mineral, sodium aluminium fluoride (Na_3AlF_6), used in the elcctrolyte reduction of ◊bauxite to aluminium. It is chiefly found in Greenland.

crystal substance with an orderly three-dimensional arrangement of its atoms or molecules, thereby creating an external surface of clearly defined smooth faces having characteristic angles between them. Examples are common salt and quartz.

covalent bond

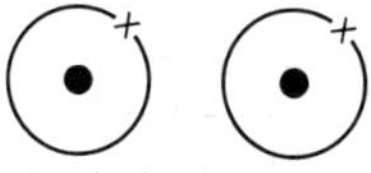

two hydrogen atoms

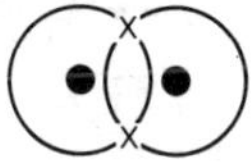

or H $\overset{\times}{\times}$ H, H–H
a molecule of hydrogen
sharing an electron pair

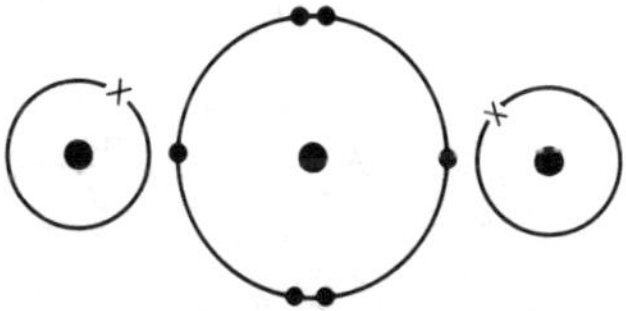

two hydrogen atoms and one
oxygen atom

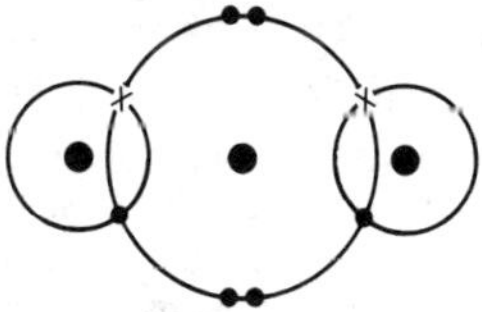

or H $\overset{\times}{\bullet}$ $\underset{\bullet\bullet}{\overset{\bullet\bullet}{O}}$ $\overset{\bullet}{\times}$ H, H–O–H
a molecule of water
showing the two covalent bonds

crystal

sodium chloride

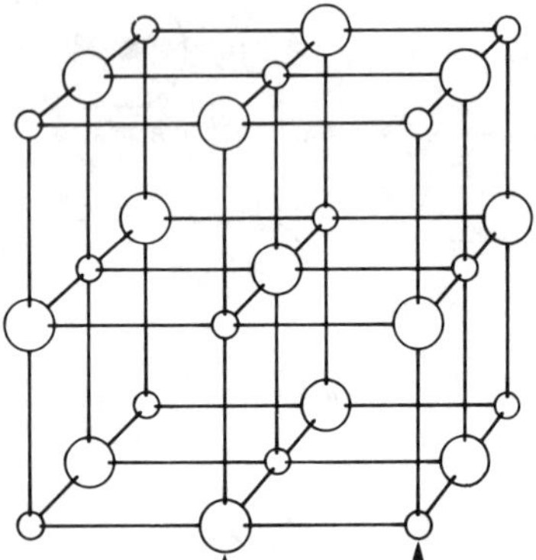

Each geometrical figure or form, many of which may be combined in one crystal, consists of two or more faces – for example, dome, prism, and pyramid.

A mineral can often be identifed by the shape of its crystals and the system of crystallization determined. A single crystal can vary in size from a sub-microscopic particle to a mass some 30 m/100 ft in length.

crystallization formation of crystals from a liquid, gas or solution.

crystallography the scientific study of crystals. In 1912, it was found that the shape and size of the unit cell of a crystal can be discovered by X-rays, thus opening up an entirely new way of 'seeing' atoms. This means of determining the atomic patterns in a crystal is known as X-ray diffraction. By this method it has been found that many substances have a unit cell that exhibits all the symmetry of the whole crystal. In table salt (sodium chloride, NaCl), for instance, the unit cell is an exact cube. It has been shown that even purified biomolecules, such as proteins and DNA, can form crystals, and such compounds may now be studied by the same method. Another field of application of X-ray analysis lies in the study of metals and alloys. Crystallography is also of use to the geologist studying rocks and soils. Many materials were not even suspected of being crystals until they were examined by X-ray crystallography.

cupronickel a copper alloy (75% copper and 25% nickel), used in hardware products and for coinage.

Curie Marie (born Sklodovska) 1867–1934. Polish scientist, who investigated radioactivity with her husband Pierre (1859–1906). They discovered radium and polonium.

Born in Warsaw, she studied in Paris from 1891. Impressed by the publication of Becquerel's experiments, Marie Curie decided to investigate the nature of uranium rays. In 1898 she reported the possible existence of some new powerful radioactive element in pitchblende ores. Her husband abandoned his own researches to assist her, and in the same year they announced the existence of polonium and radium. They isolated the pure elements in 1902.

Both scientists refused to take out a patent on their discovery and were jointly awarded the Davy Medal (1903) and the Nobel Prize for Physics (1903; with Becquerel). In 1904 Pierre was appointed to a chair in physics at the Sorbonne, and on his death in a street accident was succeeded by his wife. She wrote a *Treatise on Radioactivity* in 1910, and was awarded the Nobel Prize for Chemistry in 1911. She died a victim of radiation poisoning.

cyanamide process process used in the manufacture of calcium cyanamide. Calcium carbide is reacted with nitrogen in an electric furnace.

$$CaC_2 + N_2 \rightarrow CaCN_2 + C$$

The calcium cyanamide is used as a fertilizer under the name of Nitrolim; it reacts with water in the soil to form the ammonium ion and calcium carbonate. The ammonium is then oxidized to nitrate, which is taken up by plants.

cyanide CN^- ion derived from hydrogen cyanide (HCN), and any salt containing this ion (produced when hydrogen cyanide is neutralized by alkalis), such as potassium cyanide (KCN). The principal cyanides are potassium, sodium, calcium, mercury, gold, and copper. Certain cyanides are poisons.

cyclic compound any of a group of organic chemicals that have rings of atoms in their molecules, giving them a closed-chain structure. They may be alicyclic (for example cyclopentane), aromatic (for example benzene), or heterocylic (for example pyridine).

Alicyclic compounds have localized bonding: all the electrons are confined to their own particular bonds. Their properties are similar to those of their straight-chain counterparts.

Aromatic compounds are based on the six-carbon ◊benzene ring (C_6H_6). Some electrons in the ring have free movements between different bonds (delocalized electrons); this gives aromatic compounds specific properties that are very different to alicyclic compounds.

Heterocyclic compounds have a ring of carbon atoms with one or more carbons replaced by another element, usually nitrogen, oxygen, or sulphur. They may be alicyclic or aromatic. Pyridine is a six-membered aromatic ring with five carbons and one nitrogen(C_5H_5N).

D

Dalton John 1766–1844. British chemist, the first in modern times to propose the existence of atoms, which he considered to be the smallest parts of matter. He produced the first list of relative atomic masses in *Absorption of Gases* 1805. He was the first scientist to note and record colour blindness (he was himself colour blind).

From experiments with gases he noted that the proportions of two components combining to form another gas were always constant. From this he suggested that if substances combine in simple numerical ratios then the macroscopic weight proportions represent the relative atomic masses of those substances. He also propounded the law of partial pressures, stating that for a mixture of gases the total pressure is the sum of the pressures that would be developed by each individual gas if it were the only one present.

Davy Humphry 1778–1829. English chemist. As a laboratory assistant in Bristol in 1799, he discovered the respiratory effects of laughing gas (nitrous oxide). He discovered, by electrolysis, the elements sodium and potassium in 1807, and calcium, boron, magnesium, strontium, and barium in 1808. In addition, he established that chlorine is an element and proposed that hydrogen is present in all acids. He invented the 'safety lamp' for use in mines where methane was present, enabling the miners to work in previously unsafe conditions.

DDT abbreviation for ***dichloro-diphenyl-trichloroethane*** $(ClC_6H_4)_2CHCHCl_2$ insecticide discovered in 1939 by Swiss chemist Paul Müller. It is useful in the control of insects that spread malaria, but resistant strains develop. DDT is highly toxic and persists in the environment and in living tissue. Its use is now banned in most countries.

decomposition process whereby a chemical compound is reduced to its component substances.

decrepitation unusual features that accompany the thermal decomposition of lead(II) nitrate crystals. When these are heated, they spit and crackle and may jump out of the test tube before they decompose.

dehydrating agent substance that will remove water from another substance where the water is chemically bonded, such as water of crystallization in a hydrated salt. Powerful dehydrating agents such as concentrated sulphuric acid will even remove the elements of water from some organic compounds (for example sucrose).

dehydration removal of water from a substance to give a product with a new chemical formula; it is not the same as ◊drying.

There are two types of dehydration. For substances such as hydrated copper sulphate ($CuSO_4.5H_2O$) that contain ◊water of crystallization, dehydration means removing this water to leave the anhydrous substance. This may be achieved by heating, and is reversible.

$$CuSO_4.5H_2O \rightarrow CuSO_4 + 5H_2O$$

Some substances, such as ethanol, contain the elements of water (hydrogen and oxygen) joined in a different form. ***Dehydrating agents*** such as concentrated sulphuric acid will remove these elements in the ratio 2:1.

$$C_2H_5OH \rightarrow CH_2{=}CH_2 + H_2O$$

deliquescence the phenomenon of a substance absorbing so much moisture from the air that it ultimately dissolves in it to form a solution. Deliquescent substances make very good drying agents in the bottoms of ◊desiccators. Calcium chloride ($CaCl_2$) is one of the commonest.

density measure of the compactness of a substance; it is equal to its mass per unit volume, and is measured in kilograms per cubic metre. ***Relative density*** is the ratio of the density of a substance to that of water at 4°C.

depolarizer oxidizing agent used in dry batteries that converts hydrogen released at the negative electrode into water, preventing the build-up of gas impairing the efficiency of the battery. ◊Manganese dioxide is used for this purpose.

desalination the removal of salt, usually from sea water, to produce fresh water for irrigation or drinking. Distillation has usually been the method adopted, but in the 1970s a cheaper process, using certain polymer materials that filter the molecules of salt from the water by reverse osmosis, was developed.

desiccator airtight vessel, traditionally made of glass, in which materials may be stored either to dry them or to prevent them, once dried, from reabsorbing moisture.

The base of the desiccator is a chamber in which is placed a substance with a strong affinity for water (such as calcium chloride or silica gel), which removes water vapour from the desiccator atmosphere and from substances placed in it.

detergent surface-active cleansing agent. The common detergents are made from fats (hydrocarbons) and sulphuric acid, and their long-chain molecules have a structure similar to that of soap molecules: a salt group at one end attached to a long hydrocarbon 'tail'. They have the advantage over soap in that they do not produce scum by forming insoluble salts with the calcium and magnesium ions present in hard water.

To remove dirt, which is generally attached to materials by oil or grease, the hydrocarbon 'tails' (soluble in oil or grease) penetrate the oil or grease drops, while the 'heads' (soluble in water but insoluble in grease) remain in the water and, being salts, become ionized. Consequently the oil drops become negatively charged and tend to repel one another; thus they remain in suspension and are washed away with the dirt.

Detergents were first developed from coal tar in Germany during World War I, and synthetic organic detergents came into increasing use after World War II. Domestic powder detergents for use in hot water have alkyl benzene as their main base, and may also include bleaches and fluorescers as whiteners, perborates to free stain-removing oxygen, and water softeners. Environment-friendly detergents contain no phosphates or bleaches. Liquid detergents for washing dishes are based on ethene oxide. Cold-water detergents consists of a mixture of various alcohols, plus an ingredient for breaking down the surface tension of the water, so enabling the liquid to penetrate fibres and remove dirt. When surface-active materials escape the normal processing of sewage, they cause troublesome foam in rivers; phosphates in some detergents can also enrich the vegetation in rivers and lakes, causing ◊eutrophication.

deuterium naturally occuring heavy isotope of hydrogen, mass number 2 (one proton and one neutron), discovered by Harold Urey 1932. In nature, about one in every 6,500 hydrogen atoms is deuterium. The symbol D is sometimes used for it. Combined with oxygen, it produces 'heavy water' (D_2O), used in the nuclear industry.

deuteron nucleus of an atom of deuterium (heavy hydrogen). It consists of one proton and one neutron, and is used in the bombardment of chemical elements to synthesize other elements.

Devarda's alloy mixture of aluminium and zinc in powder form used in the test for nitrate (NO_3^-). Sodium hydroxide solution is added to the solid or solution, followed by the Devarda's alloy, and the mixture is heated. The release of ammonia gas indicates the presence of a nitrate.

diamond generally colourless, transparent mineral, the hard crystalline form of carbon. It has a ◊giant molecular structure, with its carbon molecules linked by covalent bonds to form tetrahedra. It is regarded as a precious gemstone, and is the hardest natural substance known. Industrial diamonds are used for cutting, grinding, and polishing.

diatomic molecule molecule composed of two identical atoms joined together, such as oxygen (O_2).

dibasic acid acid containing two replaceable hydrogen atoms, such as sulphuric acid (H_2SO_4). The acid can form two series of salts, the normal salt (sulphate, SO_4^{2-}) and the acid salt (hydrogensulphate HSO_4^-).

diesel oil the fuel oil used in diesel engines. Like petrol, it is a petroleum product. When used in vehicle engines, it is also known as ***derv*** – ***d****iesel-****e****ngine* ***r****oad* *v*ehicle.

diffusion one of at least three processes: the spontaneous mixing of gases or liquids (classed together as ***fluids*** in scientific usage) when brought into contact without mechanical mixing or stirring; the spontaneous passage of fluids through membranes; and the spontaneous passage of dissolved materials both through the material in which they are dissolved and also through membranes.

Diffusion can be explained by the ◊kinetic theory of matter. As the atoms or molecules of both substances are in continual random motion, over a period of time they will tend to become more evenly distributed within the total area available.

dihydroxyethane or *ethylene glycol* or *ethan-1,2-diol* CH_2OHCH_2OH thick, colourless, odourless, sweetish liquid used in antifreeze solutions. It is also used in the preparation of ethers and esters (used for explosives), as a solvent, and as a substitute for glycerine.

dilution the process of reducing the ◊concentration of a solution by the addition of ◊solvent.

The extent of a dilution normally indicates the final volume of solution required. A fivefold dilution would mean the addition of sufficient solvent to make the final volume five times the original.

dioxyethane or *diethyl ether* $C_2H_5OC_2H_5$ colourless, volatile, inflammable liquid, slightly soluble in water and miscible with ethanol. It is prepared by treatment of ethanol with excess concentrated sulphuric acid at 140°C/284°F. Dioxyethane is used as anaesthetic by vapour inhalation ('ether') and as an external cleansing agent before surgical operations. It is also used as a solvent, and in the extraction of oils, fats, waxes, resins, and alkaloids.

dipole pair of equal and opposite charges located apart, as in some ionic molecules. The product of one charge and the distance between them is the ***dipole moment***.

direct combination method of making a simple salt by heating its two constituent elements together. For example, iron(II) sulphide can be made by heating iron and sulphur, and aluminium chloride by passing chlorine over hot aluminium.

$$Fe + S \rightarrow FeS$$

$$2Al + 3Cl_2 \rightarrow 2AlCl_3$$

disinfectant agent that kills, or prevents the growth of, bacteria and other microorganisms. Chemical disinfectants include carbolic acid (phenol, used in surgery in the 1870s), ethanal, methanal, chlorine, and iodine.

dispersion the distribution of the microscopic particles of a ◊colloid. In colloidal sulphur the dispersion is the tiny particles of sulphur in the aqueous system.

displacement reaction reaction in which a less reactive element is replaced in a compound by a more reactive one. For example, the addition of powdered zinc to a solution of copper(II) sulphate displaces copper metal, which can be detected by its characteristic colour (see ◊electrochemical series).

$$Zn_{(s)} + CuSO_{4\,(aq)} \rightarrow ZnSO_{4\,(aq)} + Cu_{(s)}$$

dissociation process whereby a single compound splits into two or more smaller products that can easily recombine to form the reactant. This can be achieved in two ways.

thermal dissociation Some compounds dissociate on heating.

$$NH_4Cl_{(g)} \leftrightarrow NH_{3\,(g)} + HCl_{(g)}$$

ionization Some compounds dissociate when dissolved in water to form ions.

$$CH_3COOH + aq \leftrightarrow H^+_{(aq)} + CH_3COO^-_{(aq)}$$

In the dissociation process, a covalent bond is broken. In some instances the two portions retain their bonding electron, so no ions are formed. This usually occurs when heating is used. In a solvent such as water, the covalent bond breaks but one product retains both electrons of the bond, so forming a negative ion. The other product is therefore a positive ion.

$$A\text{–}B \rightarrow A + B$$

$$A\text{–}B \rightarrow A^+ + B\text{–}B^-$$

Where dissociation is incomplete, a ◊chemical equilibrium exists between the chemical compound and its dissociation products. The extent of incomplete dissociation is defined by a numerical value (dissociation constant).

dissolution the process of dissolving. In chemistry the term is sometimes used to describe the dissolving of a precipitate when an excess of the reagent that produced the precipitate is added.

distillation technique used to separate and purify substances that are either in solution or in a mixture of liquids that have different boiling points. The liquid is boiled; the vapours are cooled and condensed in a separate piece of apparatus (a condenser); and the liquid produced (the distillate) is collected. This form of simple distillation is used in the recovery of solvents.

Mixtures of liquids (such as ◊petroleum or aqueous ethanol) require a ◊fractionating column. In this, the mixture is boiled and the vapours enter the column. Here they condense to liquid, but as they descend they are reheated to boiling point by the hotter rising vapours. This boiling–condensing process occurs repeatedly inside the column. As the column is ascended, progressive enrichment by the lower-boiling-

distillation

laboratory apparatus for simple distillation

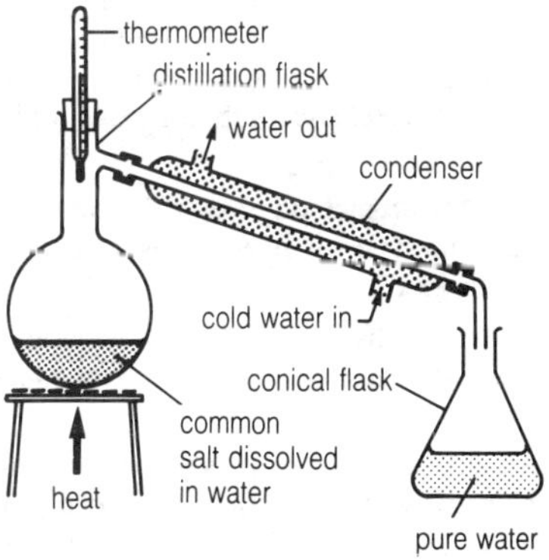

point components occurs; there is thus a temperature gradient inside the column. In crude-oil fractional distillation, groups of compounds of similar relative molecular masses and boiling points (the fractions) are tapped off from the column.

The earliest-known reference to the process is to the distillation of wine in the 12th century by Adelard of Bath. The chemical retort used for distillation was invented by Muslims, and was first seen in the West about 1570.

double bond two covalent bonds between adjacent atoms, as in the ◊alkenes (–C=C–) and ◊ketones (–C=O–).

double decomposition reaction between two chemical substances (usually ◊salts in solution) that results in the exchange of a constituent from each compound to create two different compounds.

For example, if silver nitrate solution is added to a solution of sodium chloride, there is an exchange of ions yielding sodium nitrate and silver chloride.

$$AgNO_{3\,(aq)} + NaCl_{(aq)} \rightarrow NaNO_{3\,(aq)} + AgCl_{(s)}$$

dough a mixture consisting primarily of flour, water, and yeast, which is used in the manufacture of bread. The preparation of dough involves thorough mixing (kneading) and standing in a warm place to 'prove' (increase in volume) so that the ◊enzymes in the dough can break down the starch from the flour into smaller sugar molecules, which are then fermented by the yeast. This releases carbon dioxide, which causes the dough to rise.

downward displacement method of gas collection where the gas is less dense than air, so the air in an inverted gas jar is displaced downwards by the less dense gas.

drinking water water that has been subjected to various treatments, including filtration and sterilization, to make it fit for human consumption; it is not pure water.

dry-cleaning method of cleaning textiles based on the use of volatile solvents that dissolve grease; for example, trichloroethylene. No water is used. Dry-cleaning was first developed in France in 1849.

Some solvents are known to damage the ozone layer and one, perchloroethylene, is toxic in water and gives off toxic fumes when heated.

dry ice solid carbon dioxide (CO_2), used as a refrigerant. At temperatures above –79°C, it sublimes to gaseous carbon dioxide. Water vapour in the cooled air condenses to a dense mist, and the effect is used to generate mist in the theatre.

drying removal of liquid water from a substance without altering its chemical composition (unlike ◊dehydration). Drying agents include deliquescent substances such as calcium chloride (see ◊deliquescence), concentrated acids such as sulphuric and nitric acid, and silica gel.

dye substance that, applied in solution to fabrics, imparts a colour resistant to washing. ***Direct dyes*** combine with the material of the fabric, yielding a coloured compound; ***indirect dyes*** require the presence of another substance (a mordant), with which the fabric must first be treated; ***vat dyes*** are colourless soluble substances that on exposure to air yield an insoluble coloured compound.

Naturally occurring dyes include indigo, madder (alizarin), logwood, and cochineal, but industrial dyes (introduced in the 19th century) are usually synthetic: acid green was developed 1835 and bright purple 1856. Industrial dyes include azo-dyestuffs, acridine, anthracene, and aniline.

E

efflorescence the loss of water of crystallization from crystals on standing in air, resulting in a dry powdery surface.

electrochemical series list of chemical elements arranged in descending order of the ease with which they can lose electrons to form cations (positive ions). An element can be displaced (in a ◊displacement reaction) from a compound by any element above it in the series.

electrochemistry the branch of science that studies chemical reactions involving electricity. The use of electricity to produce chemical effects, ◊electrolysis, is employed in many industrial processes, such as the manufacture of chlorine and the extraction of aluminium. The use of chemical reactions to produce electricity is the basis of batteries; see ◊cell, chemical.

Since all chemical reactions involve changes to the electron structure of atoms, all reactions are now recognized as electrochemical in nature. Oxidation, for example, was once defined as a process in which oxygen was combined with a substance, or hydrogen was removed from a compound; it is now defined as a process in which electrons are lost.

electrode conductor by which an electric current passes in or out of a substance. A positively charged electrode is called an ***anode***, because negative ions (anions) are attracted towards it; a negatively charged electrode is a ***cathode***, and cations (positive ions) are attracted towards it.

electrolysis the production of chemical changes by passing an electric current through a solution (the electrolyte), resulting in the migration of the ions to the electrodes: positive ions (cations) to the negative electrode (cathode) and negative ions (anions) to the positive electrode (anode).

During electrolysis, the ions react at the electrode, either receiving or giving up electrons. The resultant atoms may be liberated as a gas, or

electrolysis

electrolysis of acidified water

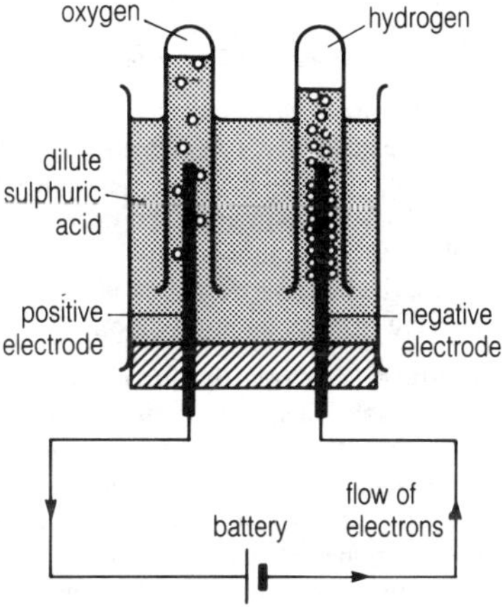

deposited as a solid on the electrode, in amounts that are proportional to the amount of current passed, as discovered by Michael Faraday.

When acidified water is electrolysed, the chemical changes that occur at the electrodes are as follows:

negative electrode: $4H^+ + 4e^- \rightarrow 2H_2$ (reduction)

positive electrode: $4OH^- - 4e^- \rightarrow 2H_2O + O_2$ (oxidation)

One application of electrolysis is ***electroplating***, in which a solution of a salt, such as silver nitrate ($AgNO_3$), is used and the object to be plated acts as the negative electrode, thus attracting silver ions (Ag^+). Electrolysis is used in many industrial processes, such as coating metals for vehicles and ships, and refining bauxite into aluminium.

electrolyte molten substance or solution in which an electric current is made to flow by the movement and discharge of ions in accordance with ◊Faraday's laws of electrolysis.

electron stable, negatively charged ◊elementary particle, a constituent of all ◊atoms and the basic particle of electricity. A beam of electrons will undergo ◊diffraction (scattering), and produce interference patterns, in the same way as ◊electromagnetic waves such as light; hence they may also be regarded as waves.

electronegativity the ease with which an atom can attract electrons to itself. Electronegative elements attract electrons, so forming negative ions. Fluorine and oxygen are the most electronegative elements; electronegativity increases along periods and decreases down groups in the ◊periodic table. In a covalent bond between two atoms of different electronegativities, the bonding electrons will be located close to the more electronegative atom, creating a ◊dipole.

electrons, delocalized ◊electrons that are not associated with individual atoms or identifiable chemical bonds, but are shared collectively by all the constituent atoms or ions of some chemical substances (such as metals, graphite, and ◊aromatic compounds).

A metallic solid consists of a three-dimensional arrangement of metal ions through which the delocalized electrons are free to travel. Aromatic compounds are characterized by the sharing of delocalized electrons by several atoms within the molecule.

electrons, localized pair of electrons in a single covalent bond that are located between the nuclei of the two contributing atoms. Such electrons cannot move beyond this area.

electroplating deposition of metals upon metallic surfaces by electrolysis for decorative and/or protective purposes. It is used in the preparation of printers' blocks, 'master' audio discs, and in many other processes.

A current is passed through a bath containing a solution of a salt of the plating metal, the object to be plated being the cathode (negative terminal); the anode (positive terminal) is either an inert substance or the plating metal. Among the metals most commonly used for plating are zinc, nickel, chromium, cadmium, copper, silver, and gold.

In ***electropolishing***, the object to be polished is made the anode in an electrolytic solution and by carefully controlling the conditions the high spots on the surface are dissolved away, leaving a high-quality stain-free surface. This technique is useful in polishing irregular stainless-steel articles.

electropositivity measure of the ability of elements (mainly metals) that donate electrons to form positive ions. The greater the metallic character, the more electropositive the element.

electrovalent bond chemical ◊bond in which the combining atoms lose or gain electrons to form ions. It is also called an ionic bond.

element substance that cannot be split chemically into simpler substances. The atoms of a particular element all have the same number of protons in their nuclei (their atomic number). Elements are classified in the ◊periodic table. Of the 109 known elements, 95 are known to occur in nature (those with atomic numbers 1–95). Eighty-one of the elements are stable; all the others, which include atomic numbers 43, 61, and from 84 up, are radioactive. Those from 96 to 109 do not occur in nature and are synthesized only, produced in particle accelerators.

Elements are classified as metals, non-metals, or semimetals depending on a combination of their physical and chemical properties; about 75% are metallic. Some elements occur abundantly (oxygen, aluminium); others occur moderately or rarely (chromium, neon); some, in particular the radioactive ones, are found in minute (neptunium, plutonium) or very minute (technetium) amounts.

Symbols (devised by Jöns Berzelius) are used to denote the elements; the symbol is usually the first letter or letters of the English or Latinized name (for example, C for carbon, Ca for calcium, Fe for iron, *ferrum*). The symbol represents one atom of the element.

elevation of boiling point raising of the boiling point of a liquid above that of the pure solvent, caused by a substance being dissolved in it. The phenomenon is observed when salt is added to boiling water; the water ceases to boil because its boiling point has been elevated.

empirical formula the simplest chemical formula for a compound. Quantitative analysis gives the proportion of each element present, and from this the empirical formula is calculated. It is related to the actual (molecular) formula by the relation:

$$(\text{empirical formula})_n = \text{molecular formula}$$

where n is a small whole number (1,2,...).

emulsion type of ◊colloid, consisting of a stable dispersion of a liquid in another liquid—for example, oil and water in some cosmetic lotions.

endothermic reaction physical or chemical change where energy is absorbed by the reactants from the surroundings. The energy absorbed is represented by the symbol $+\Delta H$. The dissolving of sodium chloride in water and the process of photosynthesis are both endothermic changes. See ◊energy of reaction.

energy the capacity for doing ◊work. ***Potential energy*** (PE) is energy deriving from position; thus a stretched spring has elastic PE; an object raised to a height above the Earth's surface, or the water in an elevated reservoir, has gravitational PE; a lump of coal and a tank of petrol, together with the oxygen needed for their combustion, have chemical PE (due to relative positions of atoms). Other sorts of PE include electrical and nuclear. Moving bodies, including atoms and molecules, possess ***kinetic energy*** (KE). Energy can be converted from one form to another, but the total quantity stays the same (in accordance with the conservation laws that govern many natural phenomena). For example,- as an apple falls, it loses gravitational PE but gains KE.

So-called energy resources are stores of convertible energy. Non-renewable resources include the fossil fuels (coal, oil, and gas) and ◊nuclear fission 'fuels' – for example, uranium–235. Renewable resources, such as wind, tidal, and geothermal power, have so far been less exploited. Hydroelectric projects are well established, and wind turbines and tidal systems are being developed. All energy sources depend ultimately on the Sun's energy.

Burning fossil fuels causes ◊acid rain and is gradually increasing the carbon dioxide content in the atmosphere, with unknown consequences for future generations. Coal-fired power stations also release significant amounts of radioactive material, and the potential dangers of nuclear power stations are greater still.

The ultimate non-renewable but almost inexhaustible energy source would be nuclear fusion (the way in which energy is generated in the

Sun), but controlled fusion is a long way off. (The hydrogen bomb is a fusion bomb.) Harnessing resources generally implies converting their energy into electrical form, because electrical energy is easy to convert to other forms and to transmit from place to place, though not to store.

energy level or ***shell*** or ***orbital*** location of the electrons in an atom. The electrons in each level have a particular energy which is dependent upon the distance of that level from the nucleus of the atom. The levels are numbered beginning with one, the nearest to the nucleus. See ◊orbital, atomic.

energy level

the arrangement of the electrons in an atom of chlorine with three energy levels (shells)

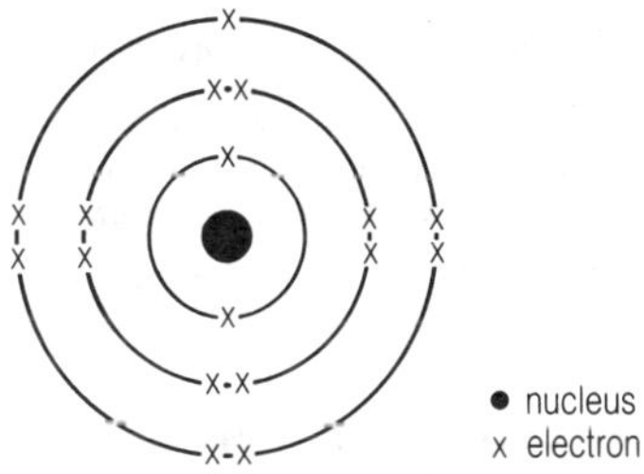

energy of reaction energy released or absorbed during a chemical reaction, also called ***enthalpy of reaction*** or ***heat of reaction***; it has the symbol ΔH.

In a chemical reaction, the energy stored in the reacting molecules is rarely the same as that stored in the product molecules. Depending on which is the greater, energy is either released (an exothermic reaction) or absorbed (an endothermic reaction) from the surroundings (see ◊conservation of energy). The amount of energy released or absorbed by the quantities of substances represented by the chemical equation is the energy of reaction.

enthalpy alternative term for ◊energy of reaction, the heat energy associated with a chemical change.

enzyme biological ◊catalyst produced in cells, capable of speeding up the chemical reactions necessary for life. Enzymes are not themselves destroyed by this process. They are large, complex ◊proteins, and are highly specific, each chemical reaction requiring its own particular enzyme. Digestive enzymes include ◊amylases (which digest starch), ◊lipases (which digest fats), and ◊proteases (which digest protein). Enzymes have many medical and industrial uses, from washing powders to drug production, and as research tools in molecular biology. They can be extracted from bacteria and moulds.

The activity and efficiency of enzymes are influenced by various factors, including temperature and pH conditions. Temperatures above 60°C damage (denature) the intricate structure of enzymes, causing reactions to cease. Each enzyme operates best within a specific pH range, and is denatured by excessive acidity or alkalinity.

Epsom salts $MgSO_4.7H_2O$ hydrated magnesium sulphate, used as a relaxant and laxative and added to baths to soothe the skin. The name is derived from a bitter saline spring at Epsom, Surrey.

equation the representation of a reaction by symbols and numbers; see ◊chemical equation.

ester organic compound formed by the reaction between an alcohol and an acid, with the elimination of water. Unlike ◊salts, esters are covalent compounds.

ethane CH_3CH_3 colourless, odourless gas, the second member of the ◊alkane series of hydrocarbons (paraffins).

ethane-1,2-diol alternative name for ◊dihydroxyethane.

ethanoate or ***acetate*** CH_3COO^- salt of ethanoic (acetic) acid. In textiles, acetate rayon is a synthetic fabric made from modified cellulose (wood pulp) treated with ethanoic acid; in photography, acetate film is a non-flammable film made of cellulose ethanoate.

ethanoic acid or ***acetic acid*** CH_3CO_2H one of the simplest ◊carboxylic acids. In the pure state it is a colourless liquid with an

ester

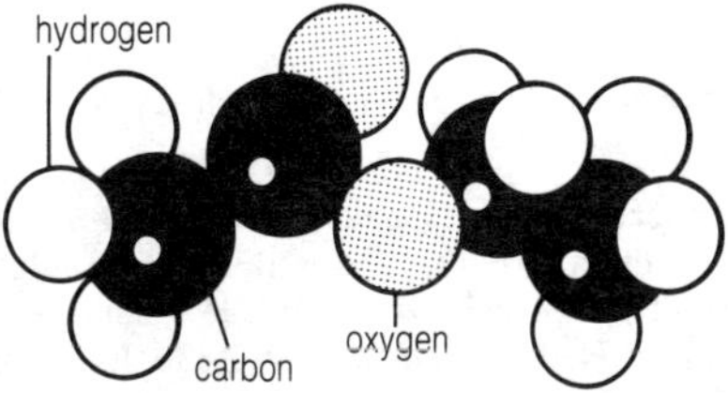

model of the ester ethyl ethanoate, CH_3 COO CH_2 CH_3

unpleasant pungent odour; it solidifies to an icelike mass of crystals at 16.7°C/62.4°F, and hence is often called ***glacial ethanoic acid***. Vinegar is 3–6% ethanoic acid.

ethanol or ***ethyl alcohol*** C_2H_5OH alcohol found in beer, wine, cider, spirits, and other alcoholic drinks. When pure, it is a colourless liquid with a pleasant odour, miscible with water or ether, and which burns in air with a pale blue flame. The vapour forms an explosive mixture with air and may be used in high-compression internal combustion engines. It is produced naturally by the fermentation of carbohydrates by yeast cells. Industrially, it can be made by absorption of ethene and subsequent reaction with water, or by the reduction of ethanal in the presence of a catalyst, and is widely used as a solvent.

ethanolic solution solution produced when a solute is dissolved in ethanol; for example, ethanolic potassium hydroxide is a solution of KOH in ethanol.

ethene or ***ethylene*** C_2H_4 colourless, flammable gas, the first member of the ◊alkene series of hydrocarbons. It is the most widely used synthetic organic chemical and is used to produce poly(ethylene) (polythene), dichloroethane, and polyvinyl chloride (PVC). It is obtained from natural gas or coal gas, or by the dehydration of ethanol.

ether any of a series of organic compounds having an oxygen atom linking the carbon atoms of two hydrocarbon radical goups (general formula R-O-R'), for example ◊dioxyethane.

ethyl alcohol alternative name for ◊ethanol.

ethylene alternative name for ◊ethene.

ethylene glycol alternative name for ◊dihydroxyethane.

ethyne or ***acetylene*** C_2H_2 colourless, inflammable gas produced by mixing calcium carbide and water. The simplest member of the ◊alkyne series of hydrocarbons, it is used in the manufacture of the synthetic rubber neoprene and in oxyacetylene welding and cutting.

Ethyne was discovered by Edmund Davy 1836. Its combustion provides more heat, relatively, than almost any other fuel known (its calorific value is five times that of hydrogen). This means that the gas gives an intensely hot flame.

eutrophication the excessive enrichment of lake waters, primarily by nitrate fertilizers, washed from the soil by rain, and by phosphates from detergents in municipal sewage. These encourage the growth of algae and bacteria which use up the oxygen in the water, thereby making it uninhabitable for fish and other animal life.

evaporation process in which a liquid turns to a vapour without its temperature reaching boiling point. A liquid left to stand in a saucer eventually evaporates because, at any time, a proportion of its molecules will be fast enough (have enough kinetic energy) to escape through the attractive intermolecular forces at the liquid surface and into the atmosphere. The rate of evaporation rises with increased temperature because as the mean kinetic energy of the liquid's molecules rises so will the number possessing enough energy to escape.

exothermic reaction reaction during which heat is given out (see ◊energy of reaction).

explosive any material capable of a sudden release of energy and the rapid formation of a large volume of gas, leading when compressed to the development of a high-pressure wave (blast).

Combustion and explosion differ essentially only in rate of reaction, and many explosives (called ***low explosives***) are capable of undergoing relatively slow combustion under suitable conditions. ***High explosives*** produce uncontrollable blasts.

The first low explosive was ◊gunpowder; the first high explosive was nitroglycerine. In 1867, Alfred ◊Nobel produced dynamite by mixing nitroglycerine with kieselguhr, a fine, chalk-like material. Other explosives now in use include trinitrotoluene (TNT); ANFO (a mixture of ammonium nitride and fuel oil), which is widely used in blasting; and pentaerythritol tetranitrate (PETN), a sensitive explosive with high power. Military explosives are often based on cyclonite (also called RDX), which is moderately sensitive but extremely powerful. Even more powerful explosives are made by mixing RDX with TNT and aluminium. ***Plastic explosives***, such as Semtex, are based on RDX mixed with oils and waxes.

The explosive force of ***atomic and hydrogen bombs*** arises from the conversion of matter to energy according to Einstein's mass–energy equation, $E = mc^2$.

F

Fahrenheit scale temperature scale invented 1714 by Gabriel Fahrenheit, no longer in scientific use. Intervals are measured in degrees (°F); °F = (°C × $^9/_5$) + 32.

Fahrenheit took as the zero point the lowest temperature he could achieve anywhere in the laboratory, and, as the other fixed point, body temperature, which he set at 96°F. On this scale, water freezes at 32°F and boils at 212°F.

faraday unit of electrical charge equal to the charge on one mole of electrons. Its value is 9.648×10^4 coulombs.

Faraday Michael 1791–1867. English chemist and physicist. In 1821 he began experimenting with electromagnetism, and ten years later discovered the induction of electric currents and made the first dynamo. He subsequently found that a magnetic field will rotate the plane of polarization of light. Faraday also investigated electrolysis.

In 1812 he began researches into electricity, and made his first electric battery. He became a laboratory assistant to Sir Humphry Davy at the Royal Institution 1813, and in 1833 succeeded him as professor of chemistry there. He delivered highly popular lectures at the Royal Institution, and published many treatises on scientific subjects. Deeply religious, he was a member of the Sandemanians (a small Congregationalist sect).

Faraday constant constant (symbol *F*) representing the electric charge carried on one mole of electrons. It is found by multiplying Avogadro's constant by the charge carried on a single electron, and is equal to 9.648×10^4 coulombs per mole. One ***faraday*** is this constant used as a unit.

The Faraday constant is used to calculate the electric charge needed to discharge a particular quantity of ions during ◊electrolysis. An ion with a single charge, such as sodium (Na^+) or chloride (Cl^-) needs one mole of electrons (one faraday of charge) to deposit one mole of atoms; a divalent ion such as calcium (Ca^{2+}) or oxygen (O^{2-}) needs two faradays; a trivalent ion such as aluminium (Al^{3+}) needs three. To calculate the number of moles of charge Q passing through an electrolyte over a period of time, the following equation is used.

$$Q = \frac{\text{current in amps} \times \text{time in seconds}}{\text{Faraday constant}}$$

fat in the broadest sense, another name for a ◊lipid: a substance that is soluble in alcohol but not water. The term is more specifically used to denote a triglyceride (a chemical containing three ◊fatty acid molecules linked to a molecule of glycerol). The three fatty acids are often of different types. Triglycerides that are liquids at room temperature are called oils; only those that are solids are called fats.

fatty acid organic compound consisting of a hydrocarbon chain, up to 24 carbon atoms long, with a carboxyl group (–COOH) at one end.

The bonds may be single or double; where a double bond occurs the carbon atoms concerned carry one instead of two hydrogen atoms. Chains with only single bonds have all the hydrogen they can carry, so they are said to to be ***saturated*** with hydrogen. Chains with one or more double bonds are said to be ***unsaturated***. Saturated fatty acids include palmitic and stearic acids; unsaturated fatty acids include oleic (one double bond), linoleic (two double bonds) and linolenic (three double bonds). Fatty acids are generally found combined with glycerol in tryglycerides or ◊fats.

Fehling's test chemical test to determine whether an organic substance is a reducing agent (substance that donates electrons to other substances in a chemical reaction).

If the test substance is heated with a freshly prepared solution containing copper(II) sulphate, sodium hydroxide and sodium potassium tartrate, the production of a brick-red precipitate indicates the presence of a reducing agent.

fermentation the breakdown of sugars by bacteria and yeasts using a method of respiration without oxygen (◊anaerobic). Fermentation processes have long been utilized in baking bread, making beer and wine, and producing cheese, yoghurt, soy sauce, and many other foodstuffs.

In baking and brewing, yeasts ferment sugars to produce ◊ethanol and carbon dioxide; the latter makes bread rise and puts bubbles into beers and champagne. Many antibiotics are produced by fermentation; it is one of the processes that can cause food spoilage.

ferric ion traditional name for the trivalent condition of iron, Fe^{3+}; the modern name is iron(III). Ferric salts are usually reddish or yellow in colour and form reddish-yellow solutions. $Fe_2(SO_4)_3$ is iron(III) sulphate (ferric sulphate).

ferro-alloy alloy of iron with a high proportion of elements such as manganese, silicon, chromium, and molybdenum. Ferro-alloys are used in the manufacture of alloy steels. Each alloy is generally named after the added metal—for example, ferrochromium.

ferrous ion traditional name for the divalent condition of iron, Fe^{2+}; the modern name is iron(II). Ferrous salts are usually green, and form yellow-green solutions. $FeSO_4$ is iron(II) sulphate (ferrous sulphate).

ferrous metal metal affected by magnetism. Iron, cobalt, and nickel are the three ferrous metals.

fertilizer substance containing a range of about 20 chemical elements necessary for healthy plant growth, used to compensate the deficiencies of poor or depleted soil. Fertilizers may be ***organic***, for example farmyard manure, composts, bonemeal, blood, and fishmeal; or ***inorganic***, in the form of compounds, mainly of nitrogen, phosphate, and potash, which have been used on a tremendously increased scale since 1945. Because the quantity applied to fields is often more than the plants need, excess fertilizers leach away to affect rivers and lakes (see ◊eutrophication).

filtrate liquid or solution that has passed through the filter paper or cloth in the filtration process.

filtration technique where suspended solid particles in a fluid are removed by passing the mixture through a porous barrier, usually paper or cloth. The particles are retained by the paper or cloth to form a

filtration

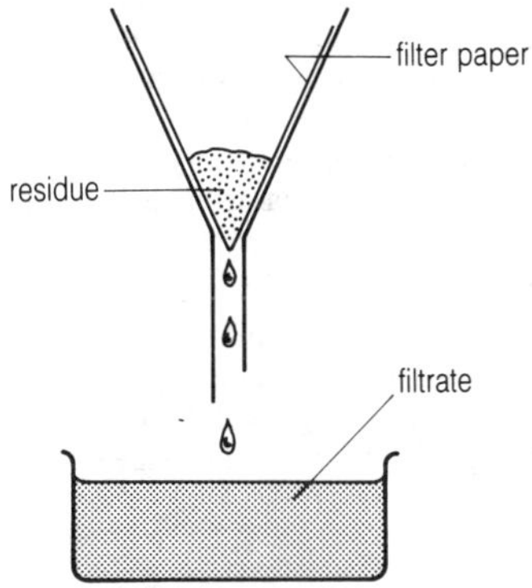

residue and the fluid passes through to make up the filtrate. Soot may be filtered from air, and suspended solids may be filtered from water.

firedamp gas that occurs in coal mines and is explosive when mixed with air in certain proportions. It consists chiefly of methane (CH_4, natural gas or marsh gas) but always contains small quantities of other gases, such as nitrogen, carbon dioxide, and hydrogen, and sometimes ethane and carbon monoxide.

fire extinguisher device for putting out a fire. Fire extinguishers work by removing one of the three conditions necessary for fire to continue (heat, oxygen, and fuel), either by cooling the fire or by excluding oxygen.

The simplest fire extinguishers contain water, which when propelled onto the fire cools it down. Water extinguishers cannot be used on electrical fires, as there is a danger of electrocution, or on burning oil, as the oil will float on the water and spread the blaze.

Many domestic extinguishers contain liquid carbon dioxide under pressure. When the handle is pressed, carbon dioxide is released as a gas that blankets the burning material and prevents oxygen reaching it.

Dry extinguishers spray powder, which then releases carbon dioxide gas. Wet extinguishers are often of the soda-acid type; when activated, sulphuric acid mixes with sodium bicarbonate, producing carbon dioxide. The gas pressure forces the solution out of a nozzle, and a foaming agent may be added to produce foam.

Some extinguishers contain halons (hydrocarbons with one or more hydrogens substituted for by a halogen such as chlorine, bromine or fluorine). These are very effective at smothering fires, but cause damage to the ◊ozone layer.

fire triangle the three essential ingredients needed to cause a fire: heat, air, and fuel. Fire-prevention strategies attempt to ensure these three conditions do not occur together. Fire control concentrates on one or more of these ingredients: water cools the temperature while foam excludes the air.

fire triangle

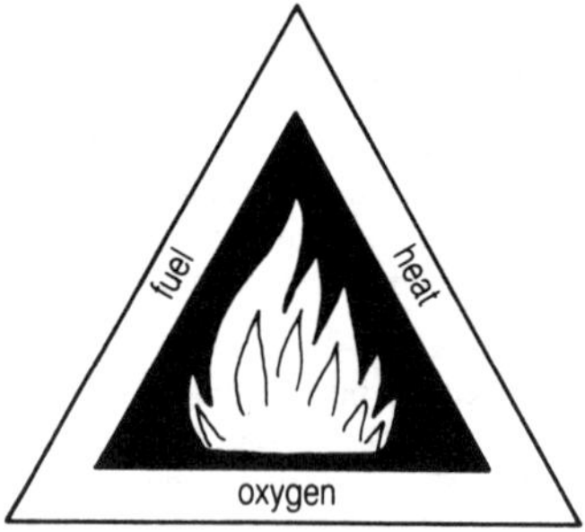

flame test the use of a flame to identify metal ◊cations present in a solid. A nichrome or platinum wire is moistened with acid, dipped in the test substance, and then held in a hot flame. The colour produced in the flame is characteristic of metals present; for example, sodium burns with a yellow flame, potassium with a lilac flame, lithium with a green flame, and calcium with a brick-red one.

flash point the lowest temperature at which a liquid or volatile solid heated under standard conditions gives off sufficient vapour to ignite on the application of a small flame.

The ***fire point*** of a material is the temperature at which full combustion occurs. For safe storage of materials such as fuel or oil, conditions must be well below the flash and fire points to reduce fire risks to a minimum.

fluid any substance, either liquid or gas, in which the molecules are relatively mobile and can 'flow'.

fluoridation addition of small amounts of fluoride salts to drinking water by certain water authorities to help prevent tooth decay. In areas where fluoride ions are naturally present in the water, research found that the incidence of tooth decay in children from those areas was reduced by more than 50%. A concentration of one part per million is sufficient to produce this beneficial effect.

fluoride F^- salt of hydrofluoric acid. Fluorides occur naturally in all water to a differing extent.

fluorine chemical element, symbol F, atomic number 9, relative atomic mass 19. It occurs naturally as the minerals fluorspar (CaF_2) and cryolite (Na_3ALF_6), and is the first member of the halogen group of elements. At ordinary temperatures it is a pale yellow, highly poisonous, and reactive gas, and it unites directly with nearly all the elements. Hydrogen fluoride is used in etching glass, and the freons, which all contain fluorine, are widely used as refrigerants.

Fluorine was discovered by the Swedish chemist Carl Wilhelm Scheele (1742–86) in 1771 and isolated by the French chemist Henri Moissan (1852–1907) in 1886. Combined with uranium as UF_6, it is used in the separation of uranium isotopes.

Minute quantities of sodium fluoride are added to some water supplies to help prevent tooth decay.

fluorocarbon compound formed by replacing the hydrogen atoms of a hydrocarbon with fluorine. Fluorocarbons are used as inert coatings, refrigerants, synthetic resins, and as propellants in aerosols.

There is concern because their release into the atmosphere depletes the ◊ozone layer, allowing more ultraviolet light from the Sun to penetrate the Earth's atmosphere, increasing the incidence of skin cancer.

formaldehyde alternative name for ◊methanal.

formalin aqueous solution of formaldehyde (methanal) used to preserve animal specimens.

formic acid alternative name for ◊methanoic acid.

formula representation of a molecule, radical, or ion, in which chemical elements are represented by their symbols. An ***empirical formula*** indicates the simplest ratio of the elements in a compound, without indicating how many of them there are or how they are combined. A ***molecular formula*** gives the number of each type of element present in one molecule. A ***structural formula*** shows the relative positions of the atoms and the bonds between them. For example, for ethanoic acid, the empirical formula is CH_2O, the molecular formula is $C_2H_4O_2$, and the structural formula is CH_3COOH. Formula is also another name for a ◊chemical equation.

fossil fuel fuel, such as coal or oil, formed from the fossilized remains of plants that lived hundreds of millions of years ago. Fossil fuels are a non-renewable resource and will run out eventually. Extraction of coal causes considerable environmental pollution, and burning coal contributes to problems of ◊acid rain and the ◊greenhouse effect.

fraction group of similar compounds, the boiling points of which fall within a particular range and which are separated during ◊fractionation.

fractionating column device in which many separate ◊distillations can occur so that a liquid mixture can be separated into its components.

Various designs exist but the primary aim is to allow maximum contact between the hot rising vapours and the cooling descending liquid. As the vapours ascend the column the mixture becomes progressively enriched in the lower boiling components, so they separate out first.

francium metallic element, symbol Fr, atomic number 87, relative atomic mass 223. It is a highly radioactive metal; the most stable isotope has a half-life of only 21 minutes. Francium was discovered by Marguérite Perey (1909–1939).

fractionating column

laboratory apparatus for fractional distillation

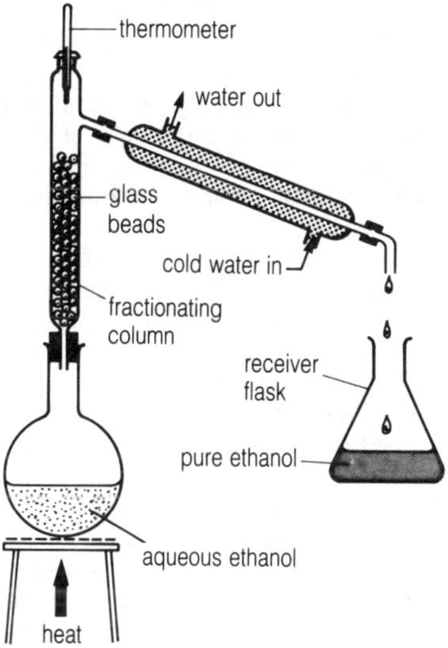

Frasch process process used to extract underground deposits of sulphur. Superheated steam is piped to the sulphur deposit and melts it. Compressed air is then pumped down to force the molten sulphur to the surface. It was developed in the USA 1891 by German-born Herman Frasch (1851–1914).

free radical an atom or molecule that has an unpaired electron and is therefore highly reactive.

Free radicals are often produced by high temperatures and are found in flames and explosions. A very simple free radical is the methyl radical CH_3 produced by the splitting of the covalent carbon-to-carbon bond in ethane.

$$CH_3CH_3 \leftrightarrow 2CH_3$$

Most free radicals are very short-lived. If free radicals are produced in living organisms they can be very damaging.

freezing change from liquid to solid state, as when water becomes ice. For a given substance, freezing occurs at a definite temperature, known as the freezing point, that is invariable under similar conditions of pressure, and the temperature remains at this point until all the liquid is frozen. The amount of heat per unit mass that has to be removed to freeze a substance is a constant for any given substance, and is known as the latent heat of fusion.

Ice is less dense than water since water expands just before its freezing point is reached. If pressure is applied, expansion is retarded and the freezing point will be lowered. The presence of dissolved substances in a liquid also lowers the freezing point (depression of freezing point), the amount of lowering being proportional to the molecular concentration of the solution. Antifreeze mixtures for car radiators and the use of salt to melt ice on roads are common applications of this principle.

Animals in arctic conditions, for example insects or fish, cope with the extreme cold either by manufacturing natural 'antifreeze' and staying active, or by allowing themselves to freeze in a controlled fashion, that is, they manufacture proteins to act as nuclei for the formation of ice crystals in areas that will not produce cellular damage, and so enable themselves to thaw back to life again.

freezing-point depression lowering of a solution's freezing point below that of the pure solvent; it depends on the number of molecules of solute dissolved in it. Thus for a single solvent, such as pure water, all substances in the same molecular concentration produce the same lowering of freezing point.

fuel any source of heat or energy, embracing the entire range of all combustibles and including anything that burns. ***Nuclear fuel*** is any material that produces energy in a nuclear reactor.

functional group small number of atoms in an arrangement that determines the chemical properties of the group and of the molecule to which it is attached (for example the carboxylic acid group –COOH, or the amine group – NH_2). Organic compounds can be considered as structural skeletons with functional groups attached.

G

Galvani Luigi 1737–1798. Italian physiologist. Born in Bologna, where he taught anatomy, he discovered galvanic, or voltaic, electricity in 1762, when investigating the contractions produced in the muscles of dead frogs by contact with pairs of different metals. His work led quickly to the invention of the electric battery by Alessandro Volta (1745–1827), and later to an understanding of how nerves control muscles.

galvanizing process for rendering iron rust-proof, by plunging it into molten zinc (the dipping method), or by electroplating it with zinc.

gas form of matter, such as air, in which the molecules move randomly in otherwise empty space, filling any size or shape of container into which the gas is put.

A sugar-lump sized cube of air at room temperature contains 30 million million million molecules moving at an average speed of 500 metres per second (1,800 kph). Gases can be liquefied by cooling, which lowers the speed of the molecules and enables attractive forces between them to bind them together.

gas collection method used to collect a gas in a laboratory preparation. The properties of the gas, and whether it is required dry, dictate the method used. Dry ammonia is collected by ◊downward displacement of air.

gas laws physical laws concerning the behaviour of gases. They include ◊Boyle's law and ◊Charles's law, which are concerned with the relationships between the pressure, temperature, and volume of an ideal (hypothetical) gas.

These laws can be combined to give the ***general*** or ***universal gas law***, which may be expressed as:

$$\frac{\text{pressure} \times \text{volume}}{\text{temperature}} = \text{constant}$$

gas collection

downward displacement

collection of a gas lighter than air

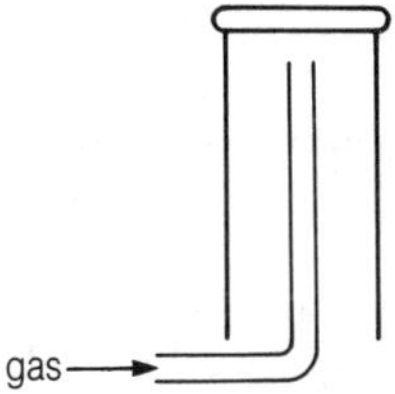

upward displacement

collection of gas heavier than air

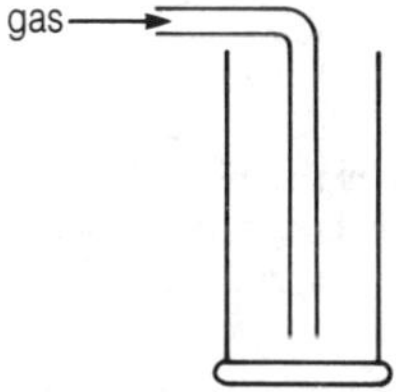

collection over water

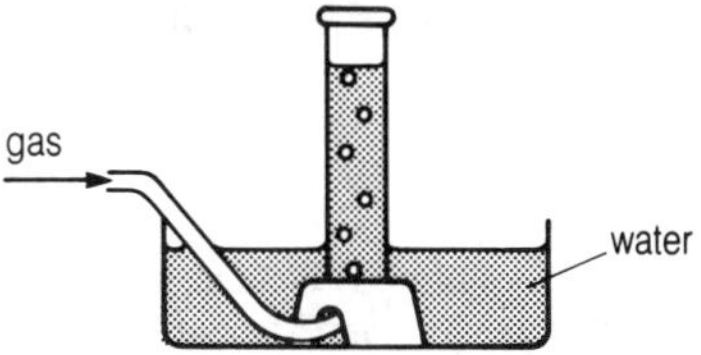

or as:

$$\frac{P_1V_1}{T_1} = \frac{P_2V_2}{T_2}$$

gas syringe graduated piece of glass apparatus used to measure accurately volumes of gases produced or consumed in a chemical reaction.

gas syringe

gel solid produced by the formation of a three-dimensional cage structure, commonly of linked large-molecular-mass polymers, in which a liquid is trapped. A gel may be a jelly-like mass (pectin, gelatin) or have a more rigid structure (silica gel).

gelatinous precipitate ◊precipitate that is viscous and jelly-like when formed; ◊aluminium hydroxide has this appearance.

germanium metallic element, symbol Ge, atomic number 32, relative atomic mass 72.6. It is a grey- white, brittle, crystalline metal in the silicon group, with chemical and physical properties between those of silicon and tin. Germanium is a semiconductor material and is used in the manufacture of transistors and integrated circuits. The oxide is transparent to infrared radiation, and is used in military applications. It was discovered 1886.

giant molecular structure or ***macromolecular structure*** solid structure made up of many similar molecules; examples include diamond, graphite, silica, and polymers.

glass brittle, usually transparent or translucent substance that is physically neither a solid nor a liquid. It is made by fusing certain types of sand (silica).

In the industrial production of common types of glass, the type of sand used, the particular chemicals added to it (for example lead, potassium, barium), and refinements of technique determine the type of glass produced. Types of glass include: soda glass; flint glass, used in cut-crystal ware; optical glass; stained glass; heat-resistant glass;

glasses that exclude certain ranges of the light spectrum; blown glass, which is either blown individually from molten glass using a tube 1.5 m long for expensive, crafted glass, or automatically blown into a mould, for example, light bulbs, and bottles; pressed glass, which is simply pressed into moulds, for jam jars, cheap vases, and light fittings; and sheet glass for windows, which is made by putting the molten glass through rollers to form a 'ribbon' or by floating molten glass on molten tin in the 'float glass' process.

Fibreglass is made from fine glass fibres. In bulk, it can be used as insulation material in construction work, or woven into material or made into glass-reinforced plastic (GRP). Fibreglass has good electrical, chemical, and weathering properties and it is also used for boat hulls, motor bodies, and aircraft components.

Glauber's salt crystalline sodium sulphate decahydrate $Na_2SO_4.10H_2O$, which melts at 31°C; the latent heat stored as it solidifies makes it a convenient thermal energy store. It is used in medicine.

glucose or ***dextrose*** $C_6H_{12}O_6$ monosaccharide sugar present in the blood, and found in honey and fruit juices. It is a source of energy for the body, being produced from other sugars and starches to form the 'energy currency' of many biochemical reactions.

It is usually prepared by hydrolysis of cane sugar or starch. Generally a yellowish syrup, it may be purified to a white crystalline powder.

glycerine common name for ***glycerol*** or ***trihydroxypropane*** $HOCH_2CH(OH)CH_2OH$ thick, colourless, odourless, sweetish liquid. It is obtained from vegetable and animal oils and fats (by treatment with acid, alkali, superheated steam, or an enzyme), or by fermentation of glucose, and is used in the manufacture of high explosives, in antifreeze solutions, to maintain moist conditions in fruits and tobacco, and in cosmetics.

glycerol alternative name for ◊glycerine.

glycol alternative name for ◊dihydroxyethane.

gold heavy, precious, yellow, metallic element, symbol Au, atomic number 79, relative atomic mass 197.0. It is unaffected by temperature changes and is highly resistant to acids. For manufacture, gold is alloyed with another strengthening metal, its purity being measured in carats on a scale of 24. In 1988 the three leading gold-producing countries were: South Africa, 621 tonnes; USA, 205 tonnes; and Australia, 152 tonnes. In 1989 gold deposits were found in Greenland with an estimated yield of 12 tonnes per year.

Gold occurs naturally in veins, but following erosion it can be transported and redeposited. It has long been valued for its durability, malleability, and ductility, and its uses include dentistry, jewellery, and electronic devices.

Graham Thomas 1805–1869. Scottish chemist who laid the foundations of physical chemistry (the branch of chemistry concerned with changes in energy during a chemical transformation) by his work on the diffusion of gases and liquids. ***Graham's Law*** states that the diffusion rate of two gases varies inversely as the square root of their densities.

His work on ◊colloids (which have larger particles than true solutions) was equally fundamental; he discovered the principle of dialysis, that colloids can be separated from solutions containing smaller molecules by the differing rates at which they pass through a semi-permeable membrane (a process he termed 'osmosis'). The human kidney uses the principle of dialysis to extract nitrogenous waste.

graphite blackish-grey, laminar, crystalline form of ◊carbon. It is used as a lubricant and as the active component of pencil lead.

The carbon atoms are strongly bonded together in sheets, but the bonds between the sheets are weak so that the sheets are free to slide over one another. Graphite has a very high melting point (3,500°C/6,332°F), which gives it mechanical strength and makes it a good conductor of heat and electricity. In its pure form it is used as a moderator in nuclear reactors.

greenhouse effect a phenomenon of the Earth's atmosphere by which solar radiation, absorbed by the Earth and re-emitted from the surface, is prevented from escaping by carbon dioxide in the air. The result is a rise in the Earth's temperature; in a garden greenhouse, the glass walls have the same effect. The concentration of carbon dioxide in the atmosphere is estimated to have risen by 25% since the Industrial Revolution, and 10% since 1950; the rate of increase is now 0.5% a year. ◊Chlorofluorocarbon levels are rising by 5% a year, and nitrous oxide levels by 0.4% a year, resulting in a global warming effect of 0.5% since 1900, and a rise of about 0.1°C a year in the temperature of the world's oceans during the 1980s. Arctic ice was 6–7 m thick in 1976 and had reduced to 4–5 m by 1987. United Nations Environment Programme estimates an increase in average world temperatures of 1.5°C/2.7°F with a consequent rise of 20 cm in sea level by 2025.

group vertical column of elements in the ◊periodic table. Elements in a group have similar physical and chemical properties; for example, the alkali metals (group I: lithium, sodium, potassium, rubidium, caesium, and francium) are all highly reactive metals that form univalent ions. There is a gradation of properties down any group: in group I, melting and boiling points decrease, and density and reactivity increase.

H

Haber Fritz 1868–1934. German chemist whose conversion of atmospheric nitrogen to ammonia opened the way for the synthetic fertilizer industry. His study of the combustion of hydrocarbons led to the commercial ◊cracking or fractionating of natural oil (petroleum) into its components (for example, diesel, petrol, and paraffin). In electrochemistry, he was the first to demonstrate that oxidation and reduction take place at the electrodes; from this he developed a general electrochemical theory.

Haber process industrial process in which ammonia is manufactured by direct combination of its elements, nitrogen and hydrogen. The reaction is carried out at 400–500°C and at 200 atmospheres pressure. The two gases, in the proportions of 1:3 by volume, are passed over a ◊catalyst of finely divided iron. Around 10% of the reactants combine, and the unused gases are recycled. The ammonia is separated by either dissolving in water or cooling to liquid.

$$N_2 + 3H_2 \leftrightarrow 2NH_3$$

haematite the principal ore of iron, consisting mainly of iron(III) (ferric) oxide, Fe_2O_3. It occurs as ***specular haematite*** (dark, metallic lustre), ***kidney ore*** (reddish radiating fibres terminating in smooth, rounded surfaces), and as a red earthy deposit.

haemoglobin protein that carries oxygen. In vertebrates it occurs in red blood cells, giving them their colour. Oxygen attaches to haemoglobin in the lungs or gills where the amount dissolved in the blood is high. This process effectively increases the amount of oxygen that can be carried in the bloodstream. The oxygen is later released in the body tissues where it is at low concentration.

half-life during ◊radioactive decay, the time in which the strength of a radioactive source decays to half its original value. It may vary from millionths of a second to billions of years.

Radioactive substances decay exponentially; thus the time taken for the first 50% of the isotope to decay will be the same as the time taken by the next 25%, and by the 12.5% after that, and so on. For example, carbon-14 takes about 5,730 years for half the material to decay; another 5,730 for half of the remaining half to decay; then 5,730 years for half of that remaining half to decay, and so on. Plutonium-239, one of the most toxic of all radioactive substances, has a half-life of about 24,000 years. In theory, the decay process is never complete and there is always some residual radioactivity. For this reason, the half-life (the time taken for 50% of the isotope to decay) is measured, rather than the total decay time.

halide the family name for a compound produced by combination of a ◊halogen, such as chlorine or iodine, with a less electronegative element (see ◊electronegativity). Halides may be formed by ionic or covalent ◊bonds.

halogen any of a group of five non-metallic elements with similar chemical bonding properties: fluorine, chlorine, bromine, iodine, and astatine. They form a linked group in the periodic table of the elements, with fluorine the most reactive and astatine the least reactive. They combine directly with most metals to form salts, for example common salt (NaCl). Each halogen has seven electrons in its valence shell, which accounts for the chemical similarities displayed by the group.

halon compound containing one or two carbon atoms, together with ◊bromine and other ◊halogens. The most commonly used halons are halon 1211 (bromochlorodifluoromethane) and halon 1301 (bromotrifluoromethane). The halons are gases and are widely used in fire extinguishers. As destroyers of the ◊ozone layer, they are up to ten times more effective than ◊chlorofluorocarbons, to which they are chemically related.

Halon levels in the atmosphere are rising by about 25% each year, mainly through the testing of fire-fighting equipment.

hardening of oils transformation of liquid oils to solid products by ◊hydrogenation. Vegetable oils contain double covalent carbon-to-carbon bonds and are thus examples of ◊unsaturated compounds.

When hydrogen is added to these double bonds, the oils become saturated. The more saturated oils are wax-like solids and are the principal constituents of margarines.

hard water water that does not lather easily with soap, and that produces 'fur' or 'scale' in kettles. It is caused by the presence of certain salts of calcium and magnesium.

Temporary hardness is caused by hydrogencarbonates. When water containing these is boiled, they are converted to insoluble carbonates that precipitate as 'scale'. ***Permanent hardness*** is caused by sulphates and silicates, which are not affected by boiling.

Water can be softened by ◊distillation, ◊ion exchange (the principle underlying commercial water softeners), addition of sodium carbonate, addition of large amounts of soap, or boiling (to remove temporary hardness). Liquid detergents readily form a lather with any water without forming a scum.

hazard labels visual system of symbols for indicating the potential dangers of handling certain substances. The symbols used are recognized internationally.

hazardous substances waste substances, usually generated by industry, which represent a hazard to the environment or to people living or working nearby. Examples include radioactive wastes, acidic resins, arsenic residues, residual hardening salts, lead, mercury, nonferrous sludges, organic solvents, and pesticides. Their economic disposal or recycling is the subject of research.

heat form of internal energy of a substance due to the kinetic energy in the motion of its molecules or atoms. Its SI unit is the joule. The extent to which a body will transfer or absorb heat (its hotness or coldness) is measured by temperature, and is related to the mean kinetic energy of its molecules.

Heat energy is transferred by conduction, convection, and radiation. Heat always flows from a region of higher temperature to one of lower temperature. Its effect on a substance may be simply to raise its temperature, cause it to expand, melt it if a solid, vaporize it if a liquid, or increase its pressure if a confined gas.

hazard labels

harmful/irritant

toxic

radioactive

explosive

flammable

corrosive

oxidizing/supports fire

Convection is the transmission of heat through a fluid (liquid or gas) on currents—for example, when the air in a room is warmed by a fire or radiator.

Conduction is the passing of heat along a medium to neighbouring parts with no visible motion accompanying the transfer of heat—for example, when the whole length of a metal rod is heated when one end is held in a fire.

Radiation is heat transfer by infrared rays. It can pass through a vacuum, travels at the same speed as light, can be reflected and refracted, and does not affect the medium through which it passes. For example, heat reaches the Earth from the Sun by radiation.

heat of reaction alternative term for ◊energy of reaction.

heavy metal metallic element of high relative atomic mass, for instance platinum, gold, and lead. Heavy metals are poisonous and tend to accumulate and persist in living systems, causing, for example, high levels of mercury (from industrial waste and toxic dumping) in shellfish and fish, which are in turn eaten by humans. Treatment of heavy-metal poisoning is difficult because available drugs are not able to distinguish between the heavy metals that are essential to living cells (zinc, copper) and those that are poisonous.

heavy water or ***deuterium oxide*** D_2O water containing the isotope deuterium instead of hydrogen (relative molecular mass 20 as opposed to 18 for ordinary water).

Its chemical properties are identical with those of ordinary water, while its physical properties differ slightly. It occurs in ordinary water in the ratio of about one part by mass of deuterium to 5,000 parts by mass of hydrogen and can be concentrated by electrolysis, the ordinary water being more readily decomposed by this means than the heavy water. In the nuclear industry, it is used as a moderator to reduce the speed of high-energy neutrons.

helium (Greek *helios* 'Sun') colourless, odourless, gaseous, nonmetallic element, symbol He, atomic number 2, relative atomic mass 4.0026. It is grouped with the ◊inert gases, is non-reactive, and forms no compounds. It is the second most abundant element (after hydrogen) in the universe, and has the lowest boiling (–268.9°C/–452°F) and melting points (–272.2°C/–458°F) of all the elements. It is present in small quantities in the Earth's atmosphere from gases issuing from radioactive elements in the Earth's crust; after hydrogen it is the second lightest element.

Helium is a component of most stars, including the Sun, where the nuclear-fusion process converts hydrogen into helium with the production of heat and light. It is obtained by compression and fractionation of naturally occurring gases and is used for inflating balloons and as a dilutant for oxygen in deep-sea breathing systems. Liquid helium is used extensively in low-temperature physics (cryogenics).

heterogeneous reaction reaction where there is an interface between the different components or reactants. Examples of heterogeneous systems are gas–solid, gas–liquid, two immiscible liquids, or different solids.

Hoffman's voltameter apparatus for collecting gases produced by the electrolysis of a liquid.

It consists of a vertical E-shaped glass tube with taps at the upper ends of the outer limbs and a reservoir at the top of the central limb. Platinum electrodes fused into the lower ends of the outer limbs are connected to a source of direct current. At the beginning of an experiment, the outer limbs are completely filled with electrolyte by opening the taps. The taps are then closed and the current switched on. Gases evolved at the electrodes bubble up the outer limbs and collect at the top, where they can be measured.

homogeneous reaction reaction in which there is no interface between the components. The term applies to all reactions where only gases are involved or where all the components are in solution.

homologous series any of a number of series of organic chemicals whose members differ by a constant relative molecular mass.

Alkanes (paraffins), alkenes (olefins), and alkynes (acetylenes) form such series whose members differ in mass by 14, 12, and 10 atomic mass units respectively. For example, the alkane homologous series begins with methane (CH_4), ethane (C_2H_6), propane (C_3H_8), butane (C_4H_{10}), and pentane (C_5H_{12}), each member differing from the previous one by a CH_2 group (or 14 atomic mass units).

homologous series

the systematic naming of simple straight-chain organic molecules

Alkane	Alcohol	Aldehyde	Ketone	Carboxylic acid	Alkene
CH_4 methane	$CH_3\,OH$ methanol	HCHO methanal	–	HCOOH methanoic acid	–
$CH_3\,CH_3$ ethane	$CH_3\,CH_2\,OH$ ethanol	$CH_3\,CHO$ ethanal	–	$CH_3\,COOH$ ethanoic acid	$CH_2\,CH_2$ ethene
$CH_3\,CH_2\,CH_3$ propane	$CH_3\,CH_2\,CH_2\,OH$ propanol	$CH_3\,CH_2\,CHO$ propanal	$CH_3\,CO\,CH_3$ propanone	$CH_3\,CH_2\,COOH$ propanoic acid	$CH_2\,CH\,CH_3$ propene
 methane	 methanol	 methanal	 propanone	 methanoic acid	 ethene

hydrate compound that has discrete water molecules combined with it. The water is known as ***water of crystallization*** and the number of water molecules associated with one molecule of the compound is denoted in both its name and chemical formula: for example, $CuSO_4.5H_2O$ is copper(II) sulphate pentahydrate.

hydration the combination of water and another substance to produce a single product. It is the opposite of ◊dehydration. For example, anhydrous copper sulphate reacts with water to give copper sulphate pentahydrate.

$$CuSO_4 + 5H_2O \rightarrow CuSO_4.5H_2O$$

hydride compound containing hydrogen and one other element only, in which the hydrogen is the more electronegative element (see ◊electronegativity).

For the more reactive metals the hydride may be an ionic compound containing a hydride anion (H^-).

hydrocarbon any of a class of compounds containing only hydrogen and carbon (for example paraffin). Hydrocarbons are obtained industrially principally from petroleum and coal tar.

hydrochloric acid HCl solution of hydrogen chloride (a colourless, acidic gas) in water. The concentrated acid is about 35% HCl and is corrosive. The acid is a typical strong, monobasic acid forming only one series of salts, the chlorides. When oxidized, for example by manganese(IV) oxide, it releases chlorine.

$$MnO_2 + 4HCl \rightarrow MnCl_2 + Cl_2 + 2H_2O$$

It has many industrial uses, including recovery of zinc from galvanized scrap iron and the production of chlorides and chlorine. It is also produced in the stomachs of animals as part of the process of digestion.

hydrocyanic acid or ***prussic acid*** HCN solution of hydrogen cyanide gas in water. It is a colourless, highly poisonous, volatile liquid, smelling of bitter almonds.

hydrogen (Greek *hydro* + *gen* 'water generator') colourless, odourless, gaseous, non-metallic element, symbol H, atomic number 1, relative atomic mass 1.00797. It is the lightest of all the elements and occurs on Earth chiefly in combination with oxygen as water. Hydrogen is the most abundant element in the universe, where it accounts for 93% of the total number of atoms and 76% total mass. It is a component of most stars, including the Sun, whose heat and light are produced through the nuclear-fusion process, which converts hydrogen into helium. When subjected to a pressure 500,000 times greater than that of the Earth's atmosphere, hydrogen becomes a solid metal. Its common and industrial uses include the hardening of fats and oils by hydrogenation and the creation of high-temperature flames for welding.

Its isotopes ◊deuterium and ◊tritium (half-life 12.5 years) are used in synthesizing elements. The name derives from the generation of water from the combustion of hydrogen; the element was named in 1787 by French chemist Louis Guyton de Morveau (1737–1816).

hydrogenation the addition of hydrogen to an unsaturated organic molecule (one that contains ◊double bonds or ◊triple bonds). The process is widely used in the manufacture of margarine and low-fat spreads by the addition of hydrogen to vegetable oils.

hydrogen bond bond between molecules that contain hydrogen covalently bonded to a more electronegative atom such as oxygen or nitrogen (see ◊electronegativity), for example water (H_2O) or ammonia (NH_3). The oxygen in water attracts electrons more than the hydrogen nucleus does, creating a small positive charge on the hydrogen and a negative charge on the oxygen (a slight ◊dipole). The opposite charges on adjacent molecules attract each other, creating the bond.

Hydrogen bonds explain certain anomolous properties of substances such as water and ammonia, for example the fact that both are liquids at room temperature although they have low relative molecular masses.

hydrogencarbonate compound containing the ion HCO_3^-, an acid salt of carbonic acid (solution of carbon dioxide in water). When heated or treated with dilute acids, they evolve carbon dioxide. The most important compounds are ◊sodium hydrogencarbonate (bicarbonate of soda) and ◊calcium hydrogencarbonate.

hydrogen cyanide HCN poisonous gas formed by the reaction of sodium cyanide with dilute sulphuric acid, used for fumigation.

The salts formed from it are cyanides—for example sodium cyanide, used in hardening steel and extracting gold and silver from their ores. If dissolved in water, hydrogen cyanide gives hydrocyanic acid.

hydrogensulphate HSO_4^- compound containing the hydrogensulphate ion. Hydrogensulphates are ◊acid salts.

hydrogen sulphide H_2S poisonous gas with the smell of rotten eggs. It is found in certain types of crude oil where it is formed by decomposition of sulphur compounds. It is removed from the oil at the refinery and the gas is converted to elemental sulphur.

hydrolysis chemical reaction in which the action of water or its ions breaks down a substance into smaller molecules. Hydrolysis occurs in certain inorganic salts in solution, in nearly all nonmetallic chlorides, in esters, and in other organic substances. It is one of the mechanisms for the breakdown of food by the body, as in the conversion of starch to glucose.

hydrophilic (Greek 'water-loving') term describing ◊functional groups with a strong affinity for water, such as the carboxylic acid group (–COOH).

If a molecule contains both a hydrophilic and a ◊hydrophobic group, it may have an affinity for both aqueous and non-aqueous molecules. Such compounds are used to stabilize an ◊emulsion or as a ◊detergent.

hydrophobic (Greek 'water-hating') term describing ◊functional groups that repel water (compare ◊hydrophilic).

hydroxide inorganic compound containing one or more hydroxyl (OH) groups, generally combined with a metal. Hydroxides include sodium hydroxide (caustic soda, NaOH), potassium hydroxide (caustic potash, KOH), and calcium hydroxide (slaked lime, $Ca(OH)_2$).

hydroxyl group an atom of hydrogen and an atom of oxygen bonded together and covalently bonded to an organic molecule. Common compounds containing hydroxyl groups are alcohols and phenols. In chemical reactions, the hydroxyl group (–OH) frequently behaves as a single entity.

hygroscopic term used to describe a substance that can absorb moisture from the air without becoming wet. There is a maximum amount of water that any particular hygroscopic substance can absorb.

I

ice solid formed by water when it freezes. It is colourless and its crystals are hexagonal. The water molecules are held together by ◊hydrogen bonds.

The freezing point, used as a standard for measuring temperature, is 0° for the Celsius scale and 32° for the Fahrenheit. Ice expands in the act of freezing (hence burst pipes), becoming less dense than water (0.9175 at 5°C/41°F).

ice

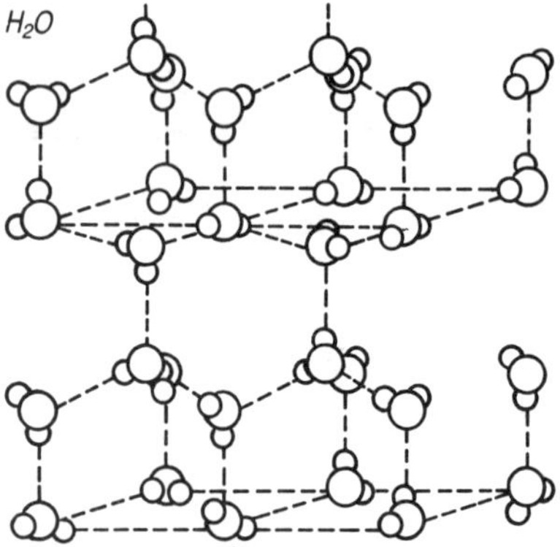

ignition temperature or ***fire point*** the minimum temperature to which a substance must be heated before it will spontaneously burn independently of the source of heat; for example, ethanol has an ignition temperature of 425°C, and a ◊flash point of 12°C.

immiscible term describing liquids that will not mix with each other, such as oil and water. When two immiscible liquids are shaken together, a turbid mixture will be produced. This normally forms separate layers on standing.

indicator compound that changes its structure and colour in response to its environment. The commonest chemical indicators detect changes in ◊pH, such as ◊litmus, or in the oxidation state of a system (redox indicators).

inert gas or ***noble gas*** element belonging to group 0 in the ◊periodic table of the elements. Inert gases are so called because they are extremely unreactive; this is because of their electronic structure. All the electron shells (◊energy levels) are full and, except for helium, they all have eight electrons in their outermost (◊valency) shell. The apparent stability of this electronic arrangement led to the formulation of the ◊octet rule to explain the different types of chemical bond found in simple compounds. In 1962 xenon was made to combine with fluorine, producing the first known compound of an inert gas.

inhibitor or ***negative catalyst*** ◊catalyst that reduces the rate of a reaction. Inhibitors are widely used in foods, medicines, and toiletries.

inorganic chemistry branch of chemistry dealing with the elements and their compounds, excluding the more complex carbon compounds which are considered in ◊organic chemistry.

The oldest known groups of inorganic compounds are ◊acids, ◊bases and ◊salts. One major group is the oxides, in which oxygen is combined with another element. Other groups are the compounds of metals with halogens (fluorine, chlorine, bromine, astatine, and iodine), which are called halides (fluorides, chlorides, and so on), and the compounds with sulphur (sulphides). The basis of the description of the elements is the ◊periodic table of elements.

In the periodic table, the elements are arranged in order of increasing atomic number (nuclear charge). The continuous sequence of elements then breaks up into 7 periods and 9 groups, the members of a group, and the sub-groups (a) and (b) into which each is divided, showing similar chemical properties. The Roman numeral at the top of each group is equal to some valency (sometimes the minimum, as in group I, sometimes the maximum, as in groups VI and VII, of the elements it contains.)

insulator any poor ◊conductor of heat, sound, or electricity. Most substances lacking free (mobile) ◊electrons, such as non-metals, are electrical or thermal insulators.

intermolecular force force of attraction between molecules, such as ◊van der Waals' force or ◊hydrogen bonding. Intermolecular forces are relatively weak, hence simple molecular compounds are gases, liquids, or low-melting-point solids.

iodide I^- salt of the ◊halide series. When a silver nitrate solution is added to any iodide solution containing dilute nitric acid, a yellow precipitate of silver iodide is formed.

$$KI_{(aq)} + AgNO_{3\,(aq)} \rightarrow AgI_{(s)} + KNO_{3\,(aq)}$$

When chlorine is passed into a solution of an iodide salt, the solution turns brown, then forms a black precipitate as iodine is produced by a ◊displacement reaction.

$$Cl_{2\,(g)} + 2I^-_{(aq)} \rightarrow 2Cl_{(aq)}{}^- + I_{2\,(s)}$$

iodine (Greek ***iodes*** 'violet') greyish-black, non-metallic element, symbol I, atomic number 53, relative atomic mass 126.9044. It is a member of the ◊halogen group. Its crystals give off, when heated, a violet vapour with an irritating odour resembling that of chlorine. It only occurs in combination with other elements. Its salts are known as iodides, which are found in sea water. As a mineral nutrient it is vital to the proper functioning of the thyroid gland, where it occurs in trace amounts as part of the hormone thyroxine. Iodine is used in photography, in medicine as an antiseptic, and in making dyes. It was discovered in 1811 by the French chemist Bernard Courtois (1777–1838).

Its radioactive isotope I-131 (half-life eight days) is a dangerous fission product from nuclear explosions and from the nuclear reactors in power plants. If ingested, it can damage the thyroid gland, by which it is taken up.

ion an atom, or group of atoms, which is either positively charged (***cation***) or negatively charged (***anion***), as a result of the loss or gain of electrons during chemical reactions or exposure to certain forms of radiation.

ion exchange process whereby the ions in one compound replace the ions in another. The exchange occurs because one of the products is insoluble in water. For example, when hard water is passed over an ion-exchange resin, the dissolved calcium and magnesium ions are replaced by either sodium or hydrogen ions, so the hardness is removed. Commercial water softeners use ion-exchange resins. The addition of ◊washing-soda crystals to hard water is also an example of ion exchange.

$$Na_2CO_{3\,(aq)} + CaSO_{4\,(aq)} \rightarrow CaCO_{3\,(s)} + Na_2SO_{4\,(aq)}$$

ion half equation equation that describes the reactions occurring at the electrodes of a chemical cell or in electrolysis. It indicates which ion is losing electrons (oxidation) or gaining electrons (reduction). Examples are given from the electrolysis of dilute hydrochloric acid (HCl).

positive electrode: $2Cl^- - 2e^- \rightarrow Cl_2$

negative electrode: $2H^+ + 2e^- \rightarrow H_2$

ionic bond or ***electrovalent bond*** bond produced when atoms of one element donate electrons to atoms of another element, forming positively and negatively charged ◊ions respectively. The electrostatic attraction between the oppositely charged ions constitutes the bond.

Each ion has the electronic structure of an inert gas (see ◊noble gas structure). The maximum number of electrons that can be gained is usually two.

ionic compound substance composed of oppositely charged ions. All salts, most bases, and some acids are examples of ionic compounds. They possess the following general properties: they are crystalline solids with a high melting point; are soluble in water and insoluble in organic solvents; and always conduct electricity when molten or in aqueous solution. A typical ionic compound is sodium chloride (Na^+Cl^-).

ionic equation equation showing only those ions in a chemical reaction that actually undergo a change, either by combining together to form an insoluble salt or by combining together to form one or more

ionic bond

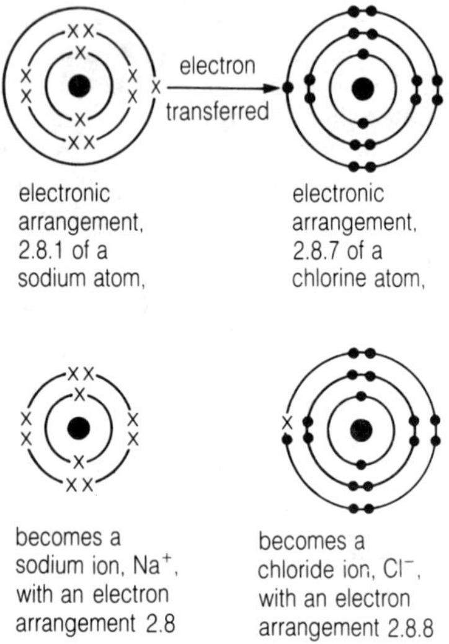

molecular compounds. Examples are the precipitation of insoluble barium sulphate when barium and sulphate ions are combined in solution, and the production of ammonia and water from ammonium hydroxide.

$$Ba^{2+}_{(aq)} + SO^{2-}_{4\,(aq)} \rightarrow BaSO_{4\,(s)}$$

$$NH^{+}_{4\,(aq)} + OH^{-}_{(aq)} \rightarrow NH_{3\,(g)} + H_2O_{(l)}$$

The other ions in the mixtures do not take part and are called ◊spectator ions.

ionization the process of ion formation. It can be achieved in two ways. The first way is by the loss or gain of electrons by atoms to form positive or negative ions.

$$Na - e^{-} \rightarrow Na^{+}$$

$$\tfrac{1}{2}Cl_2 + e^{-} \rightarrow Cl^{-}$$

In the second mechanism, ions are formed when a covalent bond breaks, as when hydrogen chloride gas is dissolved in water. One portion of the the molecule retains both electrons, forming a negative ion, and the other portion becomes positively charged. This bond-fission process is sometimes called ***dissociation***.

$$HCl_{(g)} + aq \leftrightarrow H^+_{(aq)} + Cl^-_{(aq)}$$

iron hard, malleable and ductile, silver- grey, metallic element, symbol Fe (from Latin *ferrum*), atomic number 26, relative atomic mass 55.847. It is the fourth most abundant element (the second most abundant metal, after aluminium) in the Earth's crust. Iron occurs in concentrated deposits as the ores hematite (Fe_2O_3), spathic ore ($FeCO_3$), and magnetite (Fe_3O_4). It sometimes occurs as a free metal, occasionally as fragments of iron or iron–nickel meteorites.

The metal forms two series of salts: iron(II) (ferrous) and iron(III) (ferric). The metal has the following chemical properties.

with dry air or oxygen When heated in air or oxygen, iron forms the oxide.

$$3Fe + 2O_2 \rightarrow Fe_3O_4$$

with steam When it is heated with steam in the absence of air, iron reduces the steam to hydrogen and forms the oxide.

$$3Fe + 4H_2O \rightarrow Fe_3O_4 + 4H_2$$

with dilute acids Iron forms iron(II) salts with acids, and hydrogen is evolved.

$$Fe + H_2SO_4 \rightarrow FeSO_4 + H_2$$

with air and water In moist air, iron rusts (forms hydrated iron(III) oxide).

$$4Fe + 2H_2O + 3O_2 \rightarrow 2Fe_2O_3.H_2O$$

with chlorine Iron forms iron(III) chloride when reacted with chlorine gas.

$$2Fe + 3Cl_2 \rightarrow 2FeCl_3$$

with other metals in solution Iron displaces less reactive metals when added to a solution of their salts (see ◊reactivity series).

$$Fe_{(s)} + CuSO_{4\,(aq)} \rightarrow FeSO_{4\,(aq)} + Cu_{(s)}$$

Iron is the commonest and most useful of all metals; it is strongly magnetic and is the basis for ◊steel, an alloy with carbon and other elements. In electrical equipment it is used in all permanent magnets and electromagnets, and the cores of transformers and magnetic amplifiers. See also ◊cast iron. In the human body, iron is an essential component of haemoglobin, the molecule in red blood cells that transports oxygen to all parts of the body. A deficiency in the diet causes a form of anaemia.

iron pyrites or ***pyrite*** FeS_2 common iron ore. Brassy yellow, and occurring in cubic crystals, it is often called 'fool's gold', since only those who have never seen gold would mistake it.

isomer compound having the same molecular composition and mass as another but with a different arrangement of its atoms. It has the same molecular formula but a different structural formula—for example, butane has two isomers, straight-chain and branched, C_4H_{10} and $C_2H_4(CH_3)_2$.

isoprene or ***methylbutadiene*** $CH_2CHC(CH_3)CH_2$ colourless, volatile fluid obtained from petroleum and coal, used to make synthetic rubber.

isotope one of two or more atoms that have the same atomic number (same number of protons), but which contain a different number of neutrons, thus differing in their mass numbers. They may be stable or radioactive, naturally occurring or synthesized. The term was coined by English chemist Frederick Soddy (1877–1956), a pioneer researcher in atomic disintegration.

IUPAC abbreviation for ***International Union of Pure and Applied Chemistry***, the organization that recommends the nomenclature to be used for naming substances, the units to be used, and which conventions are to be adopted when describing particular changes.

isomer

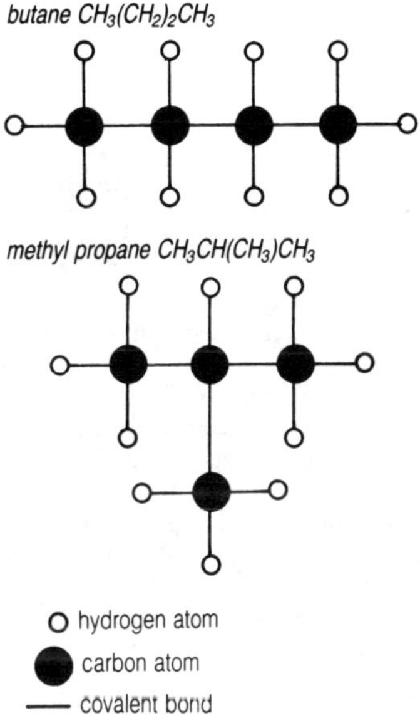
butane $CH_3(CH_2)_2CH_3$
methyl propane $CH_3CH(CH_3)CH_3$
hydrogen atom
carbon atom
covalent bond

K

kerosene thin oil obtained from the distillation of petroleum; a highly refined form is used in jet aircraft fuel. Kerosene is a mixture of hydrocarbons of the ◊alkane series.

kinetics the branch of chemistry that investigates the rates of chemical reactions.

kinetic theory theory describing the physical properties of matter in terms of the behaviour—principally movement—of its component atoms or molecules. A gas consists of rapidly moving atoms or molecules and, according to kinetic theory, it is their continual impact on the walls of the containing vessel that accounts for the pressure of the gas.

The slowing of molecular motion as temperature falls, according to kinetic theory, accounts for the physical properties of liquids and solids, culminating in the concept of no molecular motion at ◊absolute zero (0 K/–273°C). By making various assumptions about the nature of gas molecules, it is possible to derive from the kinetic theory the various ◊gas laws and ◊Avogadro's hypothesis.

krypton (Greek *kryptos* 'hidden') colourless, odourless, gaseous, non-metallic element, symbol Kr, atomic number 36, relative atomic mass 83.80. It is grouped with the inert gases and was long believed not to enter into reactions, but it is now known to combine with fluorine under certain conditions; it remains inert to all other reagents. It is present in very small quantities in the air (about 114 parts per million). It is used chiefly in fluorescent lamps, lasers, and gas-filled electronic valves.

Krypton was discovered in 1898 in the residue from liquid air by British chemists William Ramsay (1852–1916) and Morris Travers (1872–1961); the name refers to their difficulty in isolating it.

L

labelled compound or ***tagged compound*** compound in which a radioactive isotope is substituted for a stable one. Thus labelled, the path taken by the compound through a system can be followed, for example by measuring the radiation emitted. This powerful and sensitive technique is used in medicine, chemistry, biochemistry, and industry.

lactic acid or ***hydroxypropanoic acid*** $CH_3CHOHCOOH$ organic acid, a colourless, almost odourless syrup, produced by certain bacteria during fermentation. It occurs in yoghurt, buttermilk, sour cream, wine, and certain plant extracts; it is present in muscles when they are exercised hard, and also in the stomach. It is used in food preservation and in the preparation of pharmaceuticals.

lactose tasteless, white, disaccharide sugar, found in solution in milk; it forms 5% of cow's milk. Each lactose molecule is made up of two monosaccharides glucose and galactose. It is prepared commercially from the whey obtained in cheese-making.

lanthanide another name for ◊rare-earth element.

latent heat the heat that changes the state of a substance (for example, from solid to liquid) without changing its temperature.

Lavoisier Antoine Laurent 1743–1794. French chemist. He proved that combustion needed only a part of the air, which he called oxygen, thereby destroying the theory of phlogiston (an imaginary 'fire element' released during combustion). With Pierre Laplace (1749–1827), he showed that water was a compound of oxygen and hydrogen. In this way he established the basic rules of chemical combination.

lead (Indo-European *pleu* 'to flow or pour') heavy, soft, malleable, grey, metallic element, symbol Pb (from Latin *plumbum*), atomic number 82, relative atomic mass 207.19. Usually found as an ore (most often in galena), it occasionally occurs as a free metal. It is the final stable product of the decay of uranium. It is the softest and weakest of the commonly used metals, with a low melting point (hence the derivation of its name); it is a poor conductor of electricity and resists acid corrosion. Lead is a cumulative poison that enters the body from lead water pipes, lead-based paints, and leaded petrol. It is an effective shield against radiation and is used in batteries, glass, ceramics, and alloys, such as pewter and solder.

leaded petrol ◊petrol that contains ◊antiknock, a mixture of the chemicals tetraethyl lead and dibromoethane.

lead(II) nitrate $Pb(NO_3)_2$ one of only two common water-soluble compounds of lead. When heated, it decrepitates (see ◊decrepitation) and decomposes readily into oxygen, brown nitrogen(IV) oxide gas, and the red-yellow solid lead(II) oxide.

$$2Pb(NO_3)_2 \rightarrow 2PbO + 4NO_2 + O_2$$

lead(II) oxide or ***lead monoxide*** PbO yellow or red solid, an amphoteric oxide (one that reacts with both acids and bases). The other oxides of lead are the brown solid lead(IV) oxide (PbO_2) and red lead (Pb_3O_4).

Le Chatelier's principle or ***Le Chatelier-Braun principle*** the principle that if a change in conditions is imposed on a system in equilibrium, the system will react to counteract that change and restore the equilibrium.

lime or ***quicklime*** common name for ***calcium oxide*** CaO white powdery substance used in making mortar and cement and to reduce soil acidity. It is made commercially by heating calcium carbonate ($CaCO_3$) obtained from limestone or chalk. Lime readily absorbs water to become calcium hydroxide (CaOH), known as slaked lime.

lime kiln oven used to make quicklime (calcium oxide, CaO) by heating limestone (calcium carbonate, $CaCO_3$) in the absence of air. The carbon dioxide is carried away to heat other kilns and to ensure that the reversible reaction proceeds in the right direction.

$$CaCO_3 \leftrightarrow CaO + CO_2$$

limestone sedimentary rock composed chiefly of calcium carbonate $CaCO_3$, either derived from the shells of marine organisms or precipitated from solution, mostly in the ocean. Various types of limestone are used as building stone.

◊Marble is metamorphosed limestone. Certain so-called marbles are not marbles but fine-grained fossiliferous limestones that take an attractive polish. Caves commonly occur in limestone.

limewater common name for a dilute solution of slaked lime (calcium hydroxide, $Ca(OH)_2$). It is used to detect the presence of carbon dioxide.

If a gas containing carbon dioxide is bubbled through limewater, the solution turns milky owing to the formation of calcium carbonate ($CaCO_3$). Continued bubbling of the gas causes the limewater to clear again as the calcium carbonate is converted to the more soluble calcium hydrogencarbonate ($Ca(HCO_3)_2$).

lipid any of a group of organic compounds soluble in solvents such as ethanol (alcohol), but not in water. They include oils, fats, waxes, steroids, carotenoids, and other fatty substances.

liquefaction the process of converting a gas to a liquid. Liquefaction is normally associated with low temperatures and high pressures (see ◊condensation).

liquid state of matter between a ◊solid and a ◊gas. A liquid forms a level surface and assumes the shape of its container. Its atoms do not occupy fixed postions as in a crystalline solid, nor do they have freedom of movement as in a gas. Unlike a gas, a liquid is difficult to compress since pressure applied at one point is equally transmitted throughout (Pascal's principle).

liquid air air that has been cooled so much that it has liquefied. This happens at temperatures below about –196°C. The various constituent gases, including nitrogen, oxygen, argon, and neon, can be separated from liquid air by the technique of fractionation.

Air is liquefied by the ***Linde process***, in which air is alternately compressed, cooled, and expanded, the expansion resulting each time in a considerable reduction in temperature.

lithium (Greek *lithos* 'stone') soft, ductile, silver-white, metallic element, symbol Li, atomic number 3, relative atomic mass 6.941. It is one of the ◊alkali metals, has a very low density (far less than most woods), and floats on water (specific gravity 0.57); it is the lightest of all metals. Lithium is used to harden alloys, and in batteries; its compounds are used in medicine to treat manic depression.

It was named in 1818 by Swedish chemist Jöns Berzelius (1779–1848), having been discovered the previous year by his student Johan A Arfwedson (1792–1841). Berzelius named it after 'stone' because it is found in most igneous rocks and many mineral springs.

litmus dye obtained from various lichens and used as an indicator to test the acidic or alkaline nature of aqueous solutions; it turns red in the presence of acid, and blue in the presence of alkali.

lone pair pair of electrons in a bonding atomic ◊orbital that both belong to the atom itself, rather than having been paired by the sharing of electrons from different atoms. Such pairs can be involved in bonding with atoms that are deficient in electrons, forming ***dative bonds***, in which both electrons in a bond are donated by one atom.

lubricant substance used between moving surfaces to reduce friction. Carbon-based (organic) lubricants, commonly called grease and oil, are recovered from petroleum distillation.

Extensive research has been carried out on chemical additives to lubricants, which can reduce corrosive wear, prevent the accumulation of 'cold sludge' (often the result of stop-start driving in city traffic jams), keep pace with the higher working temperatures of aviation gas turbines, or provide radiation-resistant greases for nuclear power plants. Silicon-based spray-on lubricants are also used; they tend to attract dust and dirt less than carbon-based ones.

A solid lubricant is graphite, an allotropic form of carbon, either flaked or emulsified (colloidal) in water or oil.

luminous paint preparation containing a mixture of pigment, oil, and a phosphorescent sulphide, usually calcium or barium. After exposure to light it appears luminous in the dark. The luminous paint used on watch faces contains radium, which does not require exposure to light as it is radioactive.

M

macromolecule very large molecule, generally a ◊polymer.

magnesia common name for ◊magnesium oxide.

magnesium lightweight, very ductile and malleable, silver-white, metallic element, symbol Mg, atomic number 12, relative atomic mass 24.305. It is one of the ◊alkaline-earth metals, the lightest of the commonly used metals. Magnesium silicate, carbonate, and chloride are widely distributed in nature. The metal is used in alloys and flash photography. It is an necessary trace element in the human diet, and green plants cannot grow without it because it is an essential constituent of chlorophyll ($C_{55}H_{72}MgN_4O_5$).

It was named after the ancient Greek city of Magnesia, near where it was first found. It was first recognized as an element by Joseph Black (1728–1799) in 1755 and discovered in its oxide by Humphry ◊Davy in 1808.

magnesium carbonate $MgCO_3$ white solid that occurs in nature as the mineral magnesite. It is a commercial ◊antacid and the anhydrous form is used as a drying agent in table salt. When rainwater containing dissolved carbon dioxide flows over magnesite rocks, the carbonate dissolves to form magnesium hydrogencarbonate, one of the causes of temporary hardness in water.

$$H_2O + CO_2 + MgCO_3 \rightarrow Mg(HCO_3)_2$$

magnesium hydroxide $Mg(OH)_2$ white solid that occurs in nature as the mineral brucite. It is slightly soluble in water, forming an alkaline solution. It is used as an ◊antacid (milk of magnesia) and as a laxative.

magnesium oxide or ***magnesia*** MgO white powder or colourless crystals, formed when magnesium is burned in air or oxygen; a typical basic oxide. It is used to treat acidity of the stomach, and in some industrial processes – for example as a lining brick in furnaces, as it is very stable when heated (refractory oxide).

magnesium sulphate $MgSO_4$ white solid that occurs in nature in several hydrated forms. The heptahydrate $MgSO_4.7H_2O$ is known as Epsom Salts and is used as a laxative. Industrially it is used in tanning leather and in the manufacture of fertilizers and explosives.

maltose $C_{12}H_{22}O_{11}$ a ◊dissaccharide sugar in which both monosaccharide units are glucose.

It is produced by the enzymic hydrolysis of starch and is a major constituent of malt, produced in the early stages of beer and whisky manufacture.

manganese hard, brittle, grey-white, metallic element, symbol Mn, atomic number 25, relative atomic mass 54.9380. It resembles iron (and rusts), but it is not magnetic and is softer. It is used chiefly in making steel alloys, also alloys with aluminium and copper. It is used in fertilizers, paints, and industrial chemicals. It is a necessary trace element in human nutrition.

manganese(IV) oxide or ***manganese dioxide*** MnO_2 brown solid that acts as a ◊catalyst in decomposing hydrogen peroxide to obtain oxygen.

$$2H_2O_{2\,(aq)} \rightarrow 2H_2O_{(l)} + O_{2\,(g)}$$

It oxidizes concentrated hydrochloric acid to produce chlorine.

$$MnO_2 + 4HCl \rightarrow MnCl_2 + Cl_2 + 2H_2O$$

It acts as a ◊depolarizer in dry batteries by oxidizing the hydrogen gas produced to water; without this process, the performance of the battery is impaired.

marble metamorphosed ◊limestone that takes and retains a good polish; it is used in building and sculpture. In its pure form it is white and consists almost entirely of calcite $CaCO_3$. Mineral impurities give it various colours and patterns.

marsh gas gas consisting mostly of ◊methane. It is produced in swamps and marshes by the action of bacteria on dead vegetation.

mass number or ***nucleon number*** the sum (symbol *A*) of the numbers of protons and neutrons in the nucleus of an atom. It is used along with the ◊atomic number in nuclear notation and in nuclear equations (see ◊nuclear reaction); in symbols that represent nuclear isotopes, such as ${}^{14}_{6}C$, the lower number is the atomic number, and the upper number is the mass number.

melamine $C_3N_6H_6$ thermosetting ◊polymer based on urea–formaldehyde. It is extremely resistant to heat and is also scratch-resistant. Its uses include synthetic resins.

melt hot, molten mass of material. A melt of aluminium oxide in cryolite is used in the electrolytic extraction of aluminium.

melting change of state from a solid to a liquid, associated with an intake of energy (for example, if the temperature rises).

melting point the temperature at which a substance melts, or changes from a solid to liquid form. A pure substance under standard conditions of pressure (usually one atmosphere) has a definite melting point. If heat is supplied to a solid at its melting point, the temperature does not change until the melting process is complete. The melting point of ice is 0°C/32°F.

Mendeleyev Dmitri Ivanovich 1834–1907. Russian chemist who framed the periodic law 1869, which states that the chemical properties of the elements depend on their relative atomic masses. This law is the basis of the ◊periodic table of elements, in which the elements are arranged by atomic number and organized by their related groups. For his work, Mendeleyev and Lothar Meyer (who presented a similar but independent classification of the elements) received the Davy medal in 1882. From his table he predicted the properties of the then unknown elements gallium, scandium, and germanium.

meniscus the curved shape of the surface of a liquid in a thin tube, caused by the cohesive effects of surface tension. Most liquids adopt a concave curvature (viewed from above), although with highly viscous liquids (such as mercury) the meniscus is convex.

mercury or ***quicksilver*** heavy, silver-grey, metallic element, symbol Hg (from Latin *hydrargyrum*), atomic number 80, relative atomic mass

meniscus

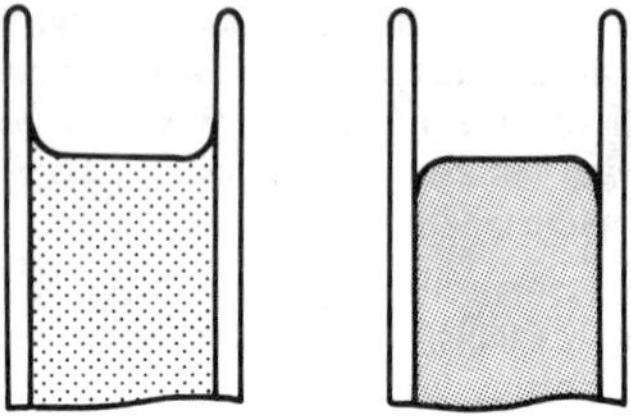

200.59. It is a dense, mobile liquid with a low melting point (–37.96°F/–38.87°C). The chief source is the mineral cinnabar (HgS). Mercury sometimes occurs as a free metal. Its alloys with other metals are called amalgams, and dentistry uses a silver–mercury amalgam for filling teeth, which may be a source of cumulative mercury poisoning to the human body. Industrial uses include drugs and chemicals, mercury-vapour lamps, arc rectifiers, power-control switches, barometers, and thermometers. Industrial dumping of this toxic substance has caused global pollution of land and waters, which has contaminated the food chain.

metal any of a class of elements with certain chemical characteristics and physical properties; they are good conductors of heat and electricity; opaque but reflect light well; malleable, which enables them to be cold-worked and rolled into sheets; and ductile, which permits them to be drawn into thin wires. Metallic elements comprise about 75% of the 109 elements shown in the ◊periodic table of the elements. They form alloys with each other, ◊bases with the hydroxyl radical (OH), and replace the hydrogen in an ◊acid to form a salt. The majority are found in the combined form only, as compounds or mineral ores; about 16 of them also occur in the elemental form, as ◊native metals. They have been put to many uses, both structural and decorative, since prehistoric times, and the Copper Age, Bronze Age, and Iron Age are named after the metal that formed the technological base for that stage of human development.

The following are widely used in commerce: ***precious metals***: gold, silver, mercury, platinum and the platinum metals, used principally in jewellery; ***heavy metals***: iron, copper, zinc, tin and lead, the common metals of engineering; ***rarer heavy metals***: nickel, cadmium, chromium, tungsten, molybdenum, manganese, cobalt, vanadium, antimony, and bismuth, used principally for alloying with the heavy metals; ***light metals***: aluminium and magnesium; ***alkali metals***: sodium, potassium and lithium; and ***alkaline-earth metals***: calcium, barium and strontium, used principally for chemical purposes.

Other metals have come to the fore because of special nuclear requirements. For example, technetium, produced in nuclear reactors, is corrosion-inhibiting; zirconium may replace aluminium and magnesium alloy in canning uranium in reactors.

metallic bond the force of attraction operating in a metal that holds the atoms together. In the metal the ◊valency electrons are able to move within the crystal and these electrons are said to be delocalized (see ◊electrons, delocalized). Their movement creates short-lived, positively charged ions. The electrostatic attraction between the delocalized electrons and the ceaselessly forming ions constitutes the metallic bond.

metallic character chemical properties associated with those elements classed as metals. These properties, which arise from the element's ability to lose electrons, are: the displacement of hydrogen from dilute acids; the formation of ◊basic oxides; the formation of ionic chlorides; and their reducing effect (see ◊reduction).

In the periodic table of the elements, metallic character increases down any group and across a period from right to left.

metalloid or ***semimetal*** element having some but not all of the properties of metals; metalloids are thus usually electrically semiconducting. They comprise the elements germanium, arsenic, antimony, and tellurium.

methanal or ***formaldehyde*** HCHO the first of the ◊aldehyde series of organic chemicals. It is a gas at ordinary temperatures, condensing to a liquid at –21°C/–5.8°F. It has a powerful penetrating smell. Dissolved in water, it is used as a biological preservative. It is used in the manufacture of plastics, dyes, foam (for example urea–formaldehyde foam, used in insulation), and in medicine.

methane CH_4 the simplest hydrocarbon of the ◊alkane (paraffin) series. Colourless, odourless, and lighter than air, it burns with a bluish flame and explodes when mixed with air or oxygen. It is the chief constituent of natural gas and also occurs in the explosive firedamp of coal mines. In marsh gas, methane forms from rotting vegetation by spontaneous combustion resulting in the pale flame seen over marshland and known as will-o'-the-wisp.

Methane is causing about 38% of the warming of the globe through the ◊greenhouse effect; the amount of methane in the air is predicted to double over the next 60 years. An estimated 15% of all methane gas released into the atmosphere is produced by cows and other cud-chewing animals.

methanoic acid common name ***formic acid*** HCOOH the first member of the ◊carboxylic acid series. It is a colourless, slightly fuming liquid that melts at 8°C/46.4°F and boils at 101°C/213.8°F. It occurs in stinging ants, nettles, sweat, and pine needles, and is used in dyeing, tanning, and electroplating.

methanol common name ***methyl alcohol*** CH_3OH the simplest of the ◊alcohols. It can be made by the dry distillation of wood (hence it is also known as wood alcohol), but is usually made from coal or natural gas. When pure, it is a colourless, flammable liquid with a pleasant odour, and is highly poisonous.

Methanol is used to produce formaldehyde (from which resins and plastics can be made), methyl-ter-butyl ether (MTB, a replacement for lead as an octane-booster in petrol), vinyl acetate (largely used in paint manufacture), and petrol.

methyl alcohol common name for ◊methanol.

methylated spirit ethanol that has been rendered undrinkable by adding poisonous substances and a violet dye; it is used for industrial purposes.

It is nevertheless drunk by some individuals, and may eventually bring about their deaths. One of the poisonous substances added to methylated spirits is ◊methanol, or methyl alcohol, and it is this that gives it its name.

methyl benzene alternative name for ◊toluene.

methyl orange $C_{14}H_{14}N_3NaO_3S$ orange-yellow powder used as an acid–base indicator in chemical tests, and as a stain in the preparation of slides of biological material. Its colour changes with pH; below pH 3.1 it is red, above pH 4.4 it is yellow.

metric system system of weights and measures developed in France in the 18th century and recognized by other countries in the 19th century. In 1960 an international conference on weights and measures recommended the universal adoption of a revised International System (Système International d'Unités, or SI), with seven prescribed 'base units': the metre (m) for length, kilogram (kg) for mass, second (s) for time, ampere (A) for electric current, kelvin (K) for thermodynamic temperature, candela (cd) for luminous intensity, and mole (mol) for quantity of matter.

Two supplementary units are included in the SI system – the radian (rad) and steradian (sr) – used to measure plane and solid angles. In addition, there are recognized derived units that can be expressed as simple products or divisions of powers of the basic units, with no other integers appearing in the expression; for example, the watt.

Some non-SI units, well established and internationally recognized, remain in use in conjunction with SI: minute, hour, and day in measuring time; multiples or submultiples of base or derived units which have long-established names, such as tonne for mass, the litre for volume; and specialist measures such as the metric carat for gemstones.

Prefixes used with metric units are tera (T) million million times; giga (G) billion (thousand million) times; mega (M) million times; kilo (k) thousand times; hecto (h) hundred times; deka (da) ten times; deci (d) tenth part; centi (c) hundreth part; milli (m) thousandth part; micro (μ) millionth part; nano (n) billionth part; pico (p) trillionth part; femto (f) quadrillionth part; atto (a) quintillionth part.

milk of magnesia common name for a suspension of magnesium hydroxide in water. It is a common ◊antacid.

mineral naturally formed inorganic substance with a particular chemical composition and an ordered internal structure. Either in their perfect crystalline form or otherwise, minerals are the constituents of rocks. In more general usage, a mineral is any substance economically valuable for mining (including coal and oil, despite their organic origins).

mineral extraction the recovery of valuable ores from the Earth's crust. The processes used include open-cast mining, shaft mining, and quarrying, as well as more specialized processes such as those used for oil and sulphur (see, for example, ◊Frasch process).

mineral oil oil obtained from mineral sources, chiefly ◊petroleum, as distinct from oil obtained from vegetable or animal sources.

mixture substance containing two or more compounds that still retain their separate physical and chemical properties. There is no chemical bonding between them and they can be separated from each other by physical means.

mobile electrons another term for delocalized electrons (see ◊electrons, delocalized). Mobile electrons are found in metals, graphite, and unsaturated molecules.

mobile ion ion that is free to move; mobile ions are only found in aqueous solution or in the ◊melt of an ◊electrolyte. The mobility of the ions in an electrolyte allows it to conduct electricity.

molarity the ◊concentration of a solution expressed as the number of ◊moles in grams of solute per litre of solution.

molar solution solution that contains one mole of a substance per litre of solvent.

molar volume volume occupied by one mole of any gas at standard temperature and pressure, equal to 2.24136×10^{-2} m^3.

mole SI unit (symbol mol) of the amount of a substance. One mole of an element that exists as single atoms weighs as many grams as its ◊atomic number (so one mole of the isotope carbon-12 weighs 12 g), and it contains 6.022045×10^{23} atoms, which is ◊Avogadro's number.

One mole of a substance is defined as the amount of that substance that contains as many elementary entities (atoms, molecules, and so on) as there are atoms in 12 g of carbon-12.

molecular formula formula indicating the actual number of atoms of each element present in a single molecule of a compound. This is determined by two pieces of information: the ◊empirical formula and the ◊relative molecular mass, which is determined experimentally.

molecular solid solid composed of molecules that are held together by relatively weak ◊intermolecular forces. Such solids have low melting points and tend to dissolve in organic solvents. Examples of molecular solids are sulphur, ice, sucrose, and solid carbon dioxide.

molecular weight another name for ◊relative molecular mass.

molecule the smallest unit of an ◊element or ◊compound that can exist and still retain the characteristics of the element or compound. A molecule of an element consists of one or more like ◊atoms; a molecule of a compound consists of two or more different atoms bonded together. They vary in size and complexity from the hydrogen molecule (H_2) to the large ◊macromolecules found in polymers. They are held together by ionic bonds, in which the atoms gain or lose electrons to form ◊ions, or covalent bonds, where electrons from each atom are shared in a new molecular orbital.

The symbolic representation of a molecule is known as its formula. The presence of more than one atom is denoted by a subscript figure – for example, one molecule of the compound water is shown as H_2O, having two atoms of hydrogen and one atom of oxygen.

According to the molecular or ◊kinetic theory of matter, molecules are in a state of constant motion, the extent of which depends on their temperature, and exert forces on one another.

Molecules were inferrable from ◊Avogadro's hypothesis 1811, but only became generally accepted in 1860 when proposed by Stanislao Cannizzarro (1826–1910).

molten state of a solid that has been heated until it melts. Hot molten solids are sometimes called a ◊melt.

molybdenum (Greek *malybdos* 'lead') heavy, hard, lustrous, silver-white, metallic element, symbol Mo, atomic number 42, relative atomic mass 95.94. The chief ore is the mineral molybdenite. The element is highly resistant to heat and conducts electricity easily. It is used in alloys, often to harden steels. It is a necessary trace element in human nutrition. It was named in 1781 by Swedish chemist Karl ◊Scheele after its isolation by P J Helm (1746–1813), for its resemblance to lead ore.

It has a melting point of 2,620°C, and is not found in the free state. As an aid to lubrication, molybdenum disulphide (MoS_2) greatly reduces surface friction between ferrous metals. Producing countries include Canada, the USA, and Norway.

monoclinic sulphur allotropic form of ◊sulphur, formed from sulphur that crystallizes above 96°C. Once formed, it very slowly changes into the rhombic allotrope.

monomer compound composed of simple molecules from which ◊polymers can be made. Under certain conditions the simple molecules (of the monomer) join together (polymerize) to form a very long chain molecule (macromolecule) called a polymer. For example, the polymerization of ethene monomers produces the polymer polyethene.

$$2nCH_2{=}CH_2 \rightarrow (CH_2{-}CH_2{-}CH_2{-}CH_2)_n$$

monosaccharide or ***simple sugar*** a ◊carbohydrate that cannot be hydrolysed (split) into smaller carbohydrate units. Examples are glucose and fructose, both of which have the molecular formula $C_6H_{12}O_6$.

mp abbreviation for ***melting point***.

multiple proportions, law of the principle that states that if two elements combine with each other to form more than one compound, then the ratio of the masses of one of them that combine with a particular mass of the other is a small whole number.

N

naphtha term originally applied to naturally occurring liquid hydrocarbons, now used for the mixtures of hydrocarbons obtained by destructive distillation of petroleum, coal tar, and shale oil. It is raw material for the petrochemical and plastics industries.

naphthalene $C_{10}H_8$ solid, white, shiny, aromatic hydrocarbon obtained from coal tar. The smell of mothballs is due to their napthalene content. It is used in making indigo and certain azo dyes, as a mild disinfectant, and an insecticide.

narcotic pain-relieving and sleep-inducing drug. The chief narcotics induce dependency, and include opium, its derivatives and synthetic modifications (such as morphine and heroin); alcohols (such as ethanol); and barbiturates.

native metal or *free metal* any of the metallic elements that occur in nature in a chemically uncombined or elemental form (in addition to any combined form). They include bismuth, cobalt, copper, gold, iridium, iron, lead, mercury, nickel, osmium, palladium, platinum, ruthenium, rhodium, tin, and silver. Some are commonly found in the free state, such as gold; others, such as mercury, occur almost exclusively in the combined state, but under unusual conditions do occur as native metals.

natural gas mixture of flammable gases found in the Earth's crust (often in association with petroleum), now one of the world's three main fossil fuels (with coal and oil). Natural gas is a mixture of ◊hydrocarbons, chiefly methane, with ethane, butane, and propane.

Before the gas is piped to storage tanks and on to consumers, butane and propane are removed and liquefied to form 'bottled gas'. Natural gas is liquefied for transport and storage, and is therefore often used where other fuels are scarce and expensive.

natural radioactivity radioactivity generated by those radioactive elements that exist in the Earth's crust. All the elements from polonium (atomic number 84) to uranium (atomic number 92) are radioactive. Radioisotopes of some lighter elements are also found in nature (for example potassium-40).

neon (Greek *neon* 'new') colourless, odourless, non-metallic, gaseous element, symbol Ne, atomic number 10, relative atomic mass 20.183. It is grouped with the ◊inert gases, is non-reactive, and forms no compounds. It occurs in small quantities in the Earth's atmosphere.

Tubes containing neon are used in electric advertising signs, giving off a fiery red glow; it is also used in lasers. Neon was discovered by Scottish chemist William Ramsay (1852–1916) and the Englishman Morris Travers (1872–1961).

neutralization process occurring when the excess acid (or excess base) in a substance is reacted with added base (or added acid) in an amount so that the resulting substance is neither acidic nor basic.

In theory neutralization involves adding acid or base as required to achieve ◊pH7.0. When the colour of an ◊indicator is used to test for neutralization, the final pH may differ from pH7.0 depending upon the indicator used.

neutral oxide oxide that has neither acidic nor basic properties (see ◊oxide). Neutral oxides are only formed by ◊non-metals. Examples are carbon monoxide, water, and nitrogen(I) oxide.

neutral solution solution of pH 7.0, in which the concentrations of $H^+_{(aq)}$ and $OH^-_{(aq)}$ ions are equal.

neutron one of the three chief subatomic particles (the others being the ◊proton and the ◊electron). Neutrons have about the same mass as protons but no electric charge, and occur in the nuclei of all ◊atoms except hydrogen. They contribute to the mass of atoms but do not affect their chemistry, which depends on the proton or electron numbers. For instance, ◊isotopes of a single element (with different masses) differ only in the number of neutrons in their nuclei and have identical chemical properties.

Outside a nucleus, a neutron is radioactive, decaying with a ◊half-life of about 12 minutes to give a proton and an electron. The neutron was discovered 1932 by the British chemist James Chadwick (1891–1974).

neutron number the number of neutrons possessed by an atomic nucleus. ◊Isotopes are atoms of the same element possessing different neutron numbers.

Nichrome trade name for a series of alloys containing mainly nickel and chromium, with small amounts of other substances such as iron, magnesium, silicon, and carbon. Nichrome has a high melting point and is resistant to corrosion. It is therefore used in electrical heating elements and as a substitute for platinum in the ◊flame test.

nickel hard, malleable and ductile, silver-white, metallic element, symbol Ni, atomic number 28, relative atomic mass 58.71. It occurs in igneous rocks and as a free metal (◊native metal), occasionally occurring in fragments of iron–nickel meteorites. It is a component of the Earth's core, which is held to consist principally of iron with some nickel. It has a high melting point, low electrical and thermal conductivity, and can be magnetized. It does not tarnish and is therefore much used for alloys, electroplating, and for coinage.

It was discovered in 1751 by Swedish mineralogist A F Cronstedt and the name given as an abbreviated form of *Kopparnickel*, Swedish for false copper, since the ore in which it is found resembles copper but yields none.

nicotine $C_{10}H_{14}N_2$ an alkaloid (nitrogenous compound) obtained from the dried leaves of the tobacco plant *Nicotiana tabacum* and used as an insecticide. A colourless oil, soluble in water, it turns brown on exposure to the air.

Nicotine in its pure form is one of the most powerful poisons known. It is named after a 16th-century French diplomat, Jacques Nicot, who introduced tobacco to France. It is the component of cigarette smoke that causes physical addiction.

nitrate salt or ester of nitric acid, containing the NO_3^- ion. Nitrates are widely used in explosives, in the chemical and pharmaceutical industries, and as fertilizers. They play a major part in the ◊nitrogen cycle. Run-off from fields results in nitrates polluting rivers and reservoirs; excess nitrate in drinking water is considered a health hazard.

nitric acid HNO_3 acid formed when nitrogen dioxide dissolves in water. It is prepared industrially by the oxidation of ammonia over a platinum catalyst.

$$4NH_3 + 5O_2 \rightarrow 4NO + 6H_2O$$

The nitrogen monoxide formed is further oxidized to nitrogen dioxide, which is then reacted with more oxygen and water to give nitric acid.

$$2NO + O_2 \rightarrow 2NO_2$$
$$4NO_2 + 2H_2O + O_2 \rightarrow 4HNO_3$$

In solution it forms fuming (95% HNO_3) or concentrated (68% HNO_3) acid. Both solutions are very corrosive and are strong oxidizing agents. They dissolve most metals, giving off brown fumes of the oxides of nitrogen and forming the nitrate salt.

$$Cu + 4HNO_3 \rightarrow Cu(NO_3)_2 + 2NO_2 + 2H_2O$$

The dilute acid generally gives off brown fumes of oxides of nitrogen when reacted with metals, showing its oxidizing power. In its other reactions it is a strong, monobasic acid.

nitrite salt or ester of nitrous acid, containing the nitrite ion (NO_2^-). Nitrites are used as preservatives (for example, to prevent the growth of botulism spores) and as colouring agents in cured meats such as bacon and sausages.

nitrogen (Greek *nitron* 'native soda', sodium or potassium nitrate) colourless, odourless, tasteless, gaseous, non-metallic element, symbol N, atomic number 7, relative atomic mass 14.0067. It forms about 78% of the Earth's atmosphere by volume and is a necessary part of all plant and animal tissues (in proteins and nucleic acids). For industrial uses it is obtained by liquifaction and fractional distillation of air.

Nitrogen has been recognized as a plant nutrient, found in manures and other organic matter, from early times, long before the complex cycle of ◊nitrogen fixation was understood. It was isolated in 1772 by

English chemist Daniel Rutherford (1749–1819) and named in 1790 by French chemist Jean Chaptal (1756–1832).

nitrogen cycle the process by which nitrogen passes through the ecosystem. Nitrogen, in the form of inorganic compounds (such as nitrates) in the soil, is absorbed by plants and turned into organic compounds (such as proteins) in plant tissue. A proportion of this nitrogen is eaten by herbivores and used for their own biological processes, with some of this in turn being passed on to the carnivores, which feed on the herbivores. The nitrogen is ultimately returned to the soil as excrement and when organisms die and are converted back to inorganic form by bacterial decomposers.

Although about 78% of the atmosphere is nitrogen, this cannot be used directly by most organisms. However, certain bacteria and cyanobacteria are capable of ◊nitrogen fixation; that is, they can extract nitrogen directly from the atmosphere and convert it to compounds such as nitrates that other organisms can use. Some nitrogen-fixing bacteria live mutually with leguminous plants (peas and beans) or other

nitrogen cycle

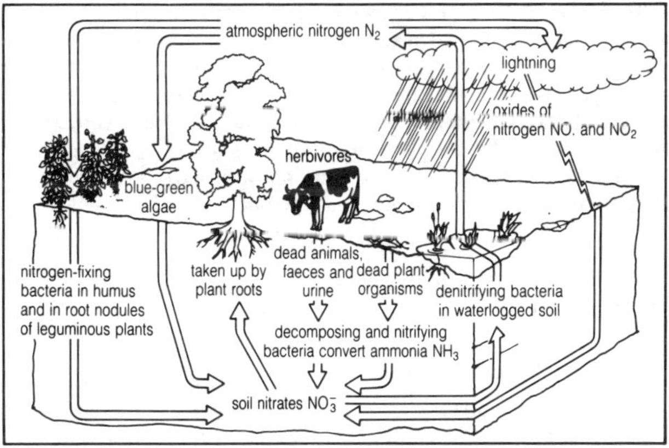

plants (for example, alder), where they form characteristic nodules on the roots. The presence of such plants increases the nitrate content, and hence the fertility, of the soil.

nitrogen fixation the process by which nitrogen in the atmosphere is converted into nitrogenous compounds by the action of microorganisms, such as cyanobacteria and bacteria, in conjunction with certain legumes. Several chemical processes duplicate nitrogen fixation to produce fertilizers; see ◊nitrogen cycle.

nitrogen oxide compound that contains only nitrogen and oxygen. All nitrogen oxides are gases, denoted by the symbol NO_x or NOX. ◊Nitrogen(II) oxide and ◊nitrogen(IV) oxide contribute to air pollution.

nitrogen(I) oxide or ***nitrous oxide*** or ***dinitrogen oxide*** N_2O colourless gas with a 'sweet' smell. It has been used as an anaesthetic, and is known familiarly as 'laughing gas'.

nitrogen(II) oxide or ***nitrogen monoxide*** NO colourless gas released when metallic copper reacts with concentrated ◊nitric acid. It is also produced when nitrogen and oxygen combine at high temperature. On contact with air it is oxidized to nitrogen(IV) oxide.

nitrogen(IV) oxide or ***nitrogen dioxide*** NO_2 brown, acidic, pungent gas that is harmful if inhaled and contributes to the formation of ◊acid rain, as it dissolves in water to form nitric acid. It is the commonest of the nitrogen oxides, and is obtained by heating most nitrate salts (for example ◊lead(II) nitrate, $Pb(NO_3)_2$). If liquefied, it gives a colourless solution (N_2O_4). It has been used in rocket fuels.

In high-temperature combustion some nitrogen and oxygen from the air combine together to form nitrogen(II) oxide.

$$N_2 + O_2 \rightarrow 2NO$$

When this oxide cools in the presence of air it is further oxidized to nitrogen(IV) oxide.

$$2NO + O_2 \rightarrow 2NO_2$$

Consequently, in notation, NO_x or NOX are used when discussing oxides of nitrogen and their emission, as both gases are present in the air. The NO_2 gas dissolves in water to give a solution of nitric acid.

nitrous acid HNO_2 weak acid that, in solution with water, decomposes quickly to form nitric acid and nitrogen dioxide.

Nobel Alfred Bernhard 1833–1896. Swedish chemist and engineer. He invented dynamite in 1867 and ballistite, a smokeless gunpowder, in 1889. He amassed a large fortune from the manufacture of explosives and the exploitation of the Baku oilfields in Azerbaijan, near the Caspian Sea, and left this fortune in trust for the endowment of five Nobel prizes.

noble gas alternative name for ◊inert gas.

noble gas structure the configuration of electrons in noble or ◊inert gases (helium, neon, argon, krypton, xenon, and radon).

This is characterized by full electron shells around the nucleus of an atom, which render the element stable. Any ion, produced by the gain or loss of electrons, that achieves an electronic configuration similar to one of the inert gases is said to have a noble gas structure.

non-conductor substance that does not conduct electricity (see ◊insulator) because of a lack of free electrons or ions. Most non-metals are non-conductors.

non-electrolyte compound that does not conduct electricity when molten or in aqueous solution. Sucrose and ethanol are examples of non-electrolytes.

non-metal one of a set of elements (around 20 in number) with certain physical and chemical properties opposite to those of metals. Non-metals accept electrons (see ◊electronegativity) and are sometimes called electronegative elements.

Their typical reactions are as follows.

with acids and alkalis Non-metals do not react with dilute acids but may react with alkalis.

$$2NaOH + Cl_2 \rightarrow NaCl + NaOCl$$

with air or oxygen They form acidic or neutral oxides.

$$S + O_2 \rightarrow SO_2$$

with chlorine They react with chlorine gas to form covalent chlorides.

$$2P_{(s)} + 3Cl_{2\,(g)} \rightarrow 2PCl_{3\,(l)}$$

with reducing agents Non-metals act as oxidizing agents.

$$2FeCl_2 + Cl_2 \rightarrow 2FeCl_3$$

Their oxides are either neutral or acidic.

NPK initials for the symbols of the three elements nitrogen, phosphorus, and potassium. These elements are essential soil nutrients for healthy crop growth. Fertilizers are made up with different amounts of these three elements to suit particular soils and crops. These initials, followed by three numbers, are seen on bags of fertilizer to indicate the relative proportion of each element in that fertilizer.

nuclear fission process whereby an atomic nucleus breaks up into two or more major fragments with the emission of two or three ◊neutrons. It is accompanied by the release of energy in the form of gamma radiation and the kinetic ◊energy of the emitted particles.

Fission occurs spontaneously in nuclei of uranium-235, the main fuel used in nuclear reactors. However, the process can also be induced by bombarding nuclei with neutrons because a nucleus that has absorbed a neutron becomes unstable and soon splits. The neutrons released spontaneously by the fission of uranium nuclei may therefore be used in turn to induce further fissions, setting up a chain reaction that must be controlled if it is not to result in a nuclear explosion.

nuclear fission

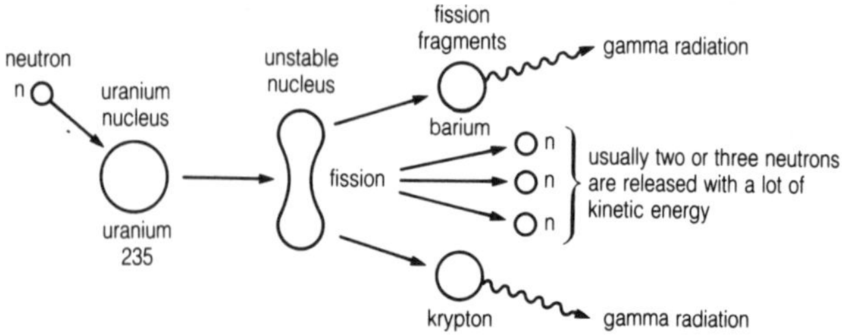

nuclear fusion process whereby two atomic nuclei are 'melted' together, or fused, with the release of a large amount of energy. Very high temperatures and pressures are thought to be required in order for the process to happen. Under these conditions the atoms involved are stripped of all their electrons so that the remaining particles, which together make up plasma, can come close together at very high speeds and overcome the mutual repulsion of the positive charges on the atomic nuclei. At very close range another nuclear force will come into play, fusing the particles together to form a larger nucleus. As fusion is accompanied by the release of large amounts of energy, the process might one day be harnessed to form the basis of commercial energy production. Methods of achieving controlled fusion are therefore the subject of research around the world.

nuclear notation method used for labelling an atom according to the composition of its nucleus. The atoms or isotopes of a particular element are represented by the symbol $^{A}_{Z}X$ where *A* is the mass number of their nuclei, *Z* is their atomic number, and *X* is the chemical symbol for that element.

This notation provides a convenient method for describing all types of nuclear reactions.

nuclear reaction reaction involving the nuclei of atoms. Atomic nuclei can undergo changes either as a result of radioactive decay, as in the decay of radium to radon (with the emission of an alpha particle) or as a result of particle bombardment in a machine or device, as in the production of cobalt-60 by the bombardment of cobalt-59 with neutrons.

$$^{226}_{88}\mathrm{Ra} \rightarrow {}^{222}_{86}\mathrm{Rn} + {}^{4}_{2}\mathrm{He}$$

$$^{59}_{27}\mathrm{Co} + {}^{1}_{0}\mathrm{n} \rightarrow {}^{60}_{27}\mathrm{C0} + \gamma \text{ (gamma radiation)}$$

◊Nuclear fission and ◊nuclear fusion are examples of nuclear reactions. The enormous amounts of energy released may be explained by the mass–energy relation put forward by Einstein, stating that $E = mc^2$ (where *E* is energy, *m* is mass, and *c* is the velocity of light).

The sum of the masses of the products of a nuclear reaction is less than the sum of the masses of the reacting particles. This lost mass is converted to energy according to Einstein's equation.

nuclear safety the use of nuclear energy has given rise to concern over safety. Anxiety has been heightened by accidents such as Windscale (UK), Five Mile Island (USA) and Chernobyl (USSR). There has also been mounting concern about the production and disposal of ***nuclear waste***, the radioactive and toxic by-products of the nuclear-energy and nuclear-weapons industries. Burial on land or at sea, raises problems of safety, environmental pollution, and security. Nuclear waste may have an active life of several thousand years and there are no guarantees of the safety of the various methods of disposal. Nuclear safety is still a controversial subject. In 1990 a scientific study revealed an increased risk of leukemia in children whose fathers had worked at Sellafield between 1950 and 1985. Sellafield (UK) is the world's greatest discharger of radioactive waste.

nucleic acid complex organic acid made up of a long chain of nucleotides. The two types, known as DNA (deoxyribonucleic acid) and RNA (ribonucleic acid), form the basis of heredity. The nucleotides are made up of a sugar (deoxyribose or ribose), a phosphate group, and one of four purine or pyrimidine bases. The order of the bases along the nucleic acid strand contains the genetic code.

nucleon any particle present in the atomic nucleus. ◊Protons and ◊neutrons are nucleons.

nucleon number alternative name for the ◊mass number of an atom.

nucleus the positively charged central part of an ◊atom, which constitutes almost all its mass. Except for hydrogen nuclei, which have only ◊protons, nuclei are composed of both protons and ◊neutrons. Surrounding the nuclei are ◊electrons, which contain a negative charge equal to the protons, thus giving the atom a neutral charge.

nylon synthetic long-chain polymer similar in chemical structure to protein. Nylon was the first all-synthesized fibre, made from petroleum, natural gas, air, and water by the Du Pont firm in 1938. It is used in the manufacture of moulded articles, textiles, and medical sutures. Nylon fibres are stronger and more elastic than silk and are relatively insensitive to moisture and mildew. Nylon is used for hosiery and woven goods, simulating other materials such as silks and furs; it is also used for carpets.

O

octane rating a numerical classification of petroleum fuels indicating their combustion characteristics.

The efficient running of an internal combustion engine depends on the ignition of a petrol–air mixture at the correct time during the cycle of the engine. Higher-rated petrol burns faster than lower-rated fuels. The use of the correct grade must be matched to the engine.

The numerical value of the octane rating is calculated as the percentage of iso-octane (2-methylpentane) in a mixture with heptane that has similar combustion properties to the fuel.

octet rule rule stating that elements combine in a way that gives them the electronic structure of the nearest ◊inert gas. All the inert gases except helium have eight electrons in their outermost shell, hence the term octet is used.

This rule is helpful in understanding how the two principal types of bonding – ionic and covalent – are formed.

oil inflammable substance, usually insoluble in water, and chiefly composed of carbon and hydrogen. Oils may be solids (fats and waxes) or liquids. The three main types are: ***essential oils***, obtained from plants; ***fixed oils***, obtained from animals and plants; and ***mineral oils***, obtained chiefly from the refining of ◊petroleum, or crude oil.

Essential oils are volatile liquids that have the odour of their plant source and are used in perfumes, flavouring essences, and in aromatherapy. Fixed oils are mixtures of ◊esters of fatty acids, of varying consistency, found in both animals (for example, fish oils) and plants (in nuts and seeds). They are used as food; to make soaps, paints, and varnishes; and for lubrication.

olefin common name for ◊alkene.

orbital, atomic the region around the nucleus of an atom (or, in a molecule, around several nuclei) in which an ◊electron is most likely to be found. According to quantum theory, the position of an electron is uncertain; it may be found at any point. However, it is more likely to be found in some places than in others, and it is these that make up the orbital.

An atom or molecule has numerous orbitals, each of which has a fixed size and shape. An orbital is characterized by three numbers, called ***quantum numbers***, representing its energy (and hence size), its angular momentum (and hence shape), and its orientation. Each orbital can be occupied by one or (if their spins are aligned in opposite directions) two electrons.

ore body of rock, a vein within it, or a deposit of sediment, worth mining for the economically valuable mineral it contains.

The term is usually applied to sources of metals. Occasionally metals are found uncombined (◊native metals), but more often they occur as compounds such as carbonates, sulphides, or oxides. The ores often contain unwanted impurities that must be removed when the metal is extracted. Commercially valuable ores include bauxite (aluminium oxide, Al_2O_3) haematite (iron(III) oxide, Fe_2O_3), zinc blende (zinc sulphide, ZnS), and rutile (titanium dioxide, TiO_2).

organic chemistry branch of chemistry that deals with carbon compounds, in particular the more complex ones. Organic compounds form the chemical basis of life and are more abundant than inorganic compounds. The basis of organic chemistry is the ability of carbon to form long chains of atoms, branching chains, rings, and other complex structures. In a typical organic compound, each carbon atom forms a bond with each of its neighbouring carbon atoms in the chain or ring, and two more with hydrogen atoms (carbon has a valency of four). Other atoms that may be involved in organic molecules include oxygen and nitrogen. Compounds containing only carbon and hydrogen are known as ***hydrocarbons***.

Organic chemistry is largely the chemistry of a great variety of homologous series – those in which the molecular formulae, when arranged in ascending order, form an arithmetical progression. The physical properties undergo a gradual change from one member to the next.

organic chemistry

common organic molecule groupings

formula	*name*	*structural formula*
CH_3	methyl	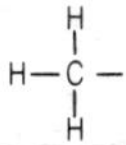
CH_2CH_3	ethyl	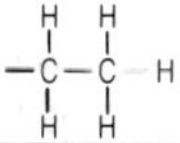
CC	double bond	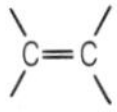
CHO	aldehyde	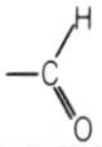
CH_2OH	alcohol	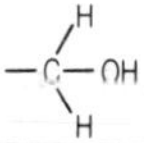
CO	ketone	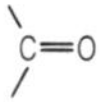
COOH	acid	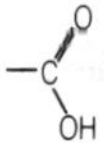
CH_2NH_2	amino	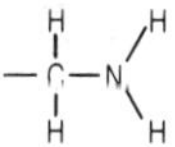
C_6H_6	benzene ring	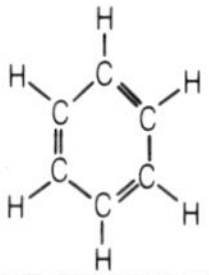

The chain of carbon atoms forming the backbone of an organic molecule may be built up from beginning to end without branching; or it may throw off branches at one or more points. This division of organic compounds is known as the ***open-chain*** or ***aliphatic*** compounds. Sometimes, however, the ropes of carbon atoms curl round and form rings. These constitute the second division of organic compounds, known as ***closed-chain, ring***, or ***cyclic*** compounds. Other structural varieties are known.

Many organic compounds are made only by living organisms (for example proteins, carbohydrates), and it was once believed organic compounds could not be made by any other means. This was disproved when Wöhler synthesized urea, but the name 'organic' (that is 'living') chemistry has remained in use. Many organic compounds are derived from oil, which represents the chemical remains of millions of microscopic marine organisms.

In inorganic chemistry, a specific formula usually represents one substance only, but in organic chemistry, it is exceptional for a molecular formula to represent only one substance. Substances having the same molecular formula are called ***isomers***, and the relationship is known as ***isomerism***.

Hydrocarbons form one of the most prolific of the many organic types; fuel oils are largely made up of hydrocarbons. Typical groups containing only carbon, hydrogen, and oxygen are alcohols, aldehydes, ketones, ethers, esters, and carbohydrates. Among groups containing nitrogen are amides, amines, nitro-compounds, amino-acids, proteins, purines, alkaloids, and many others, both natural and artificial. Other organic types contain sulphur, phosphorus, or halogen elements.

The most fundamental of all natural processes are oxidation, reduction, hydrolysis, condensation, polymerization, and molecular rearrangement. In nature, such changes are often brought about through the agency of promoters known as ***enzymes***, which act as catalytic agents in promoting specific reactions. The most fundamental of all natural processes is ***synthesis***, or building up. In living organisms, the energy stored in carbohydrate molecules, derived originally from sunlight, is released by slow oxidation and utilized by the organisms. The complex carbohydrates thereby revert to carbon dioxide and water, from where they were built up with absorption of energy. Thus, a so-called carbon food cycle exists in nature. In a corresponding nitrogen food cycle, complex proteins are synthesized in nature from carbon dioxide, water, soil nitrates, and ammonium salts, and these proteins ultimately revert to the elementary raw materials from which they came, with the discharge of their energy of chemical combination.

osmosis the movement of solvent (liquid) through a semipermeable membrane separating solutions of different concentrations. The solvent passes from the more dilute solution to the more concentrated solution until the two concentrations are equal. Applying external pressure to the solution on the more concentrated side arrests osmosis, and is a measure of the osmotic pressure of the solution.

Many cell membranes behave as semipermeable membranes, and osmosis is a vital mechanism in the transport of fluids in living organisms – for example in the transport of water from the roots up the stems of plants.

osmosis

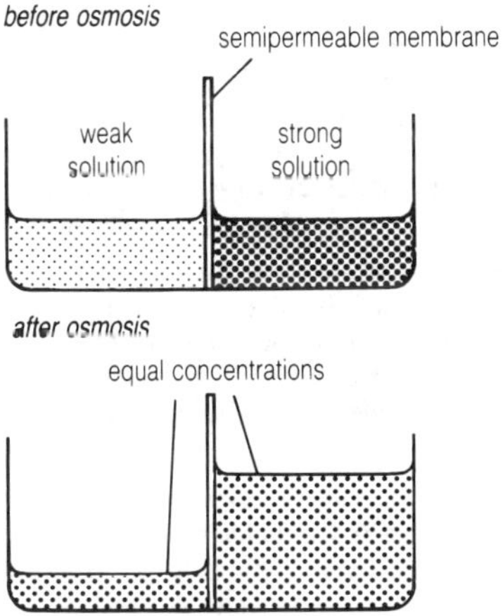

oxalic acid $(COOH)_2.2H_2O$ white, poisonous solid, soluble in water, alcohol, and ether. Oxalic acid is found in rhubarb, and its salts (oxalates) occur in wood sorrel (genus *Oxalis*, family Oxalidaceae) and other plants. It is used in the leather and textile industries, in dyeing and bleaching, ink manufacture, metal polishes, and for removing rust and ink stains.

It is one of the oldest known organic acids.

oxidation the loss of ◊electrons, gain of oxygen, or loss of hydrogen by an atom, ion, or molecule during a chemical reaction.

Oxidation may be brought about by reaction with another compound (oxidizing agent), which simultaneously undergoes ◊reduction, or electrically at the anode (positive terminal) of an electric cell.

oxidation number Roman numeral often seen in a chemical name, indicating the ◊valency of the element immediately before the number. Examples are lead(II) nitrate, manganese(IV) oxide, and potassium manganate(VII).

oxide compound of oxygen and another element, frequently produced by burning the element or a compound of it in air or oxygen.

Oxides of metals are normally ◊bases and will react with an acid to produce a ◊salt in which the metal forms the cation (positive ion). Some of them will also react with a strong alkali to produce a salt in which the metal is part of a complex anion (negative ion; see ◊amphoteric). Most oxides of non-metals are acidic (dissolve in water to form an ◊acid). Some oxides display no pronounced acidic or basic properties.

oxide film thin film of oxide formed on the surface of some metals as soon as they are exposed to the air. This oxide film makes the metal much more resistant to a chemical attack. The considerable lack of reactivity of aluminium to most reagents arises from this propoerty. The thickness of the oxide film can be increased by ◊anodizing the aluminium.

oxidizing agent substance that will oxidize another substance (see ◊oxidation). In a redox reaction, the oxidizing agent is the substance that is itself reduced. Common oxidizing agents include oxygen, chlorine, nitric acid, and potassium manganate(VII).

oxyacetylene torch gas torch that burns ethene (acetylene) in pure oxygen, producing a high-temperature (3,000°C/5,400°F) flame. It is widely used in welding to fuse metals. In a cutting torch, a jet of oxygen burns through metal already melted by the flame.

oxygen (Greek *oxys* 'acid' *genes* 'forming') colourless, odourless, tasteless, non-metallic, gaseous element, symbol O, atomic number 8, relative atomic mass 15.9994. It is the most abundant element in the Earth's crust (almost 50% by mass), forms about 21% by volume of the atmosphere, and is present in combined form in water, carbon dioxide, silicon dioxide (quartz), iron ore, calcium carbonate (limestone) and many other substances. Life on Earth evolved using oxygen, which is a by-product of ◊photosynthesis and the basis for ◊respiration in plants and animals. ◊Ozone is an allotrope of oxygen.

Oxygen is very reactive and combines with all other elements except the ◊inert gases and fluorine. In nature it exists as a molecule composed of two atoms (O_2); single atoms of oxygen are very short-lived owing to their reactivity. They can be produced in electric sparks and by the Sun's ultraviolet radiation in space, where they rapidly combine with molecular oxygen to form ozone.

Oxygen was first identified by English chemist Joseph ◊Priestley in 1774 and independently in the same year by Swedish chemist Karl ◊Scheele. It was named in 1777 by French chemist Antoine ◊Lavoisier.

ozone O_3 highly reactive pale-blue gas with a penetrating odour. Ozone is an allotrope of oxygen (see ◊allotropy), made up of three atoms of oxygen. It is formed when the molecule of the stable form of oxygen (O_2) is split by ultraviolet radiation or electrical discharge. It forms a layer in the upper atmosphere, which protects life on Earth from ultraviolet rays, a cause of skin cancer. At lower levels it contributes to the ◊greenhouse effect.

A continent-sized hole has formed over Antarctica as a result of damage to the ozone layer caused by ◊chlorofluorocarbons. In 1989 ozone depletion was 50% over the Antarctic compared with 3% over the Arctic.

At ground level, ozone can cause asthma attacks, stunted growth in plants, and corrosion of certain materials. It is produced by the action of sunlight on car exhaust fumes, and is a major air pollutant in hot summers. The US Environment Protection Agency recommends people should not be exposed for more than one hour a day to ozone levels of 120 parts per billion (ppb), while the World Health Organization recommends a lower 76–100 ppb. It is known that even at levels of 60 ppb ozone causes respiratory problems, and may cause the yields of some crops to fall. In the USA, the annual economic loss due to ozone has been estimated at $5.4 billion. Ozone is a powerful oxidizing agent and is used industrially in bleaching

P

palladium lightweight, ductile and malleable, silver-white, metallic element, symbol Pd, atomic number 46, relative atomic mass 106.4. It is one of the so-called platinum group of metals, and is resistant to tarnish and corrosion. It often occurs in nature as a free metal in a natural alloy with platinum. Palladium is used as a catalyst, in alloys of gold (to make white gold) and silver, in electroplating, and in dentistry.

It was discovered 1803 by British physicist William Wollaston (1766–1828), and named after the then recently discovered asteroid Pallas (found in 1802).

paraffin common name for ◊alkane, any member of the series of hydrocarbons with the general formula C_nH_{2n}+2. The lower members are gases, such as methane (marsh or natural gas). The middle ones (mainly liquid) form the basis of petrol, kerosene, and lubricating oils, while the higher ones (paraffin waxes) are used in ointment and cosmetic bases.

The fuel commonly sold as paraffin in Britain is more correctly called kerosene.

particle size the size of the grains that make up a powder. The grain size has an effect on certain properties of a substance. Finely divided powders have a greater surface area for contact; they therefore react more quickly, dissolve more readily, and are of increased efficiency as catalysts compared with their larger-sized counterparts.

pentanol common name ***amyl alcohol*** $C_5H_{11}OH$ clear, colourless, oily liquid, usually having a characteristic choking odour. It is obtained by the fermentation of starches and from the distillation of petroleum.

peptide molecule comprising two or more ◊amino acid molecules (not necessarily different) joined by ***peptide bonds***, whereby the acid group of one acid is linked to the amino group of the other

(–CO–NH–). The number of amino acid molecules in the peptide is indicated by referring to it as a di-, tri-, or polypeptide (two, three, or many amino acids).

Proteins are built up of interacting polypeptide chains with various types of bonds occurring between the chains. Incomplete hydrolysis (splitting up) of a protein yields a mixture of peptides, examination of which helps to determine the sequence in which the amino acids occur within the protein.

peptide bond bond that joins two peptides together within a protein. The carboxyl (–COOH) group on one ◊amino acid reacts with the amino ($-NH_2$) group on another amino acid to form a peptide bond (–CO–NH–) with the elimination of water.

period horizontal row of elements in the ◊periodic table. There is a gradation of properties along each period, from metallic (group I, the alkali metals) to non-metallic (group VII, the halogens).

periodic table of the elements classification of the elements following the statement by ◊Mendeleyev 1869 that 'the properties of elements are in periodic dependence upon their atomic weight'. (Today elements are classified by their atomic numbers rather than by their relative molecular masses.) There are striking similarities in the chemical properties of the elements in each of the vertical columns (called ***groups***), which are numbered I–VII, and a gradation of properties along the horizontal rows (called ***periods***). These features are a direct consequence of the electronic (and nuclear) structure of the atoms of

periodic table: schematic view

group number	I	II	III	IV	V	VI	VII	0
number of electrons in outermost shell	1	2	3	4	5	6	7	8
valencies	1	2	3	4	5(3)	6(2)	7(1)	0
formula of the chloride of element M	MCl	MCl_2	MCl_3	MCl_4	MCl_5 (MCl_3)	MCl_2	MCl	–
formula of the oxide of element M	M_2O	MO	M_2O_3	MO_2	M_2O_5 (M_2O_3)	MO_3	M_2O_7 (M_2O)	–

the elements. The periodic table summarizes the major properties of the elements and how they change, and enables predictions to be made. The full version is printed in Appendix I.

Metallic character increases down a group and across a period from right to left. The group number (I–VII) indicates the number of electrons in the outermost shell and hence the maximum ◊valency. The formulae of simple compounds can thus be deduced.

permanent hardness hardness of water that cannot be removed by boiling (see ◊hard water).

Perspex technical name ***polymethylmethacrylate*** (PMMA) trademark for a clear, lightweight, tough plastic first produced 1930. It is widely used for watch glasses, advertising signs, domestic baths, motorboat windshields, aircraft canopies, and protective shields. It is manufactured under other names: Plexiglas (in the USA), Oroglas (in some European countries), and Lucite.

petrochemical chemical derived from the processing of ◊petroleum. The ***petrochemical industry*** is a term embracing those industrial manufacturing processes that obtain their raw materials from the processing of petroleum.

petrol mixture of hydrocarbons derived from petroleum, mainly used as a fuel for internal combustion engines. It is colourless and highly volatile. In the USA, petrol is called gasoline.

Leaded petrol contains antiknock (a mixture of tetraethyl lead and dibromoethane), which improves the combustion of petrol and the performance of a car engine. The lead from the exhaust fumes enters the atmosphere, mostly as simple lead compounds. In recent years the level of lead in the air has risen, and there is strong evidence that it can act as a nerve poison on young children and can cause mental impairment. This has prompted a gradual switch to the use of ***unleaded petrol*** in the UK, which gained momentum owing to a change in the tax on petrol in 1989 that made it cheaper to buy unleaded fuel. Unleaded petrol contains a different mixture of hydrocarbons, and has a lower ◊octane rating than leaded petrol.

petroleum or ***crude oil*** natural mineral oil, a thick greenish-brown flammable liquid found underground in permeable rocks. Petroleum

petroleum

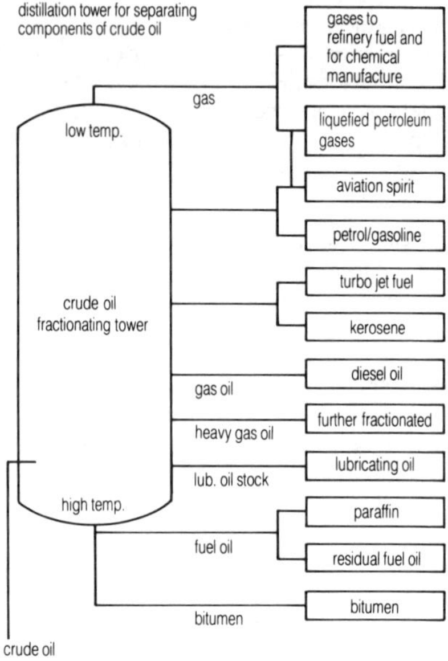

consists of hydrocarbons mixed with oxygen, sulphur, nitrogen, and other elements in varying proportions. It is thought to be derived from ancient organic material that has been converted by, first, bacterial action, then heat and pressure (but its origin may be chemical also). From crude petroleum, various products are made by distillation and other processes; for example, fuel oil, petrol (gasoline), kerosene, diesel, lubricating oil, paraffin wax, and petroleum jelly.

The organic material in petroleum was laid down millions of years ago (hence, fossil fuel). Petroleum is often found as large underground lakes floating on water but under a layer of ◊natural gas (mainly methane), trapped below layers of rock that do not allow it to pass through. Oil may flow naturally from wells under gas pressure from above or water pressure from below, causing it to rise up the borehole, but many oil wells require pumping to bring the oil to the surface.

Petroleum products and chemicals are used in large quantities in the manufacture of detergents, artificial fibres, plastics, insecticides, fertilizers, pharmaceuticals, toiletries, and synthetic rubber. Aviation fuel is a volatile form of petrol.

The burning of fuels derived from petroleum is a major cause of air pollution. Its transport can lead to major catastrophes—for example, the *Torrey Canyon* tanker lost off SW England 1967, which led to an agreement by the international oil companies 1968 to pay compensation for massive shore pollution. The 1989 oil spill in Alaska from the *Exxon Valdez* damaged the area's fragile environment, despite clean-up efforts. Drilling for petroleum involves the risks of accidental spillage and drilling-rig accidents. The problems associated with petroleum have led to the various alternative energy technologies.

pH scale for measuring acidity or alkalinity. A pH of 7.0 (distilled water) indicates neutrality, below 7 is acid, while above 7 is alkaline. The pH value of a solution equals the negative logarithm of the concentration of hydrogen ions.

The scale runs from 0 to 14. Strong acids, as used in car batteries, have a pH of about 2; acidic fruits such as citrus fruits are about pH 4. Fertile soils have a pH of about 6.5 to 7.0, while weak alkalis such as soap are 9 to 10. Corrosive alkalis such as concentrated sodium or potassium hydroxide (lye) are pH 13.

The pH of a solution can be measured by using a broad-range indicator, either in solution or in the form of a paper strip. The colour produced by the indicator is compared with a colour code related to the pH value. An alternative method is to use a pH meter fitted with a glass electrode.

phase physical state of matter: for example, ice and liquid water are different phases of water; a mixture of the two is termed a two-phase system.

pH

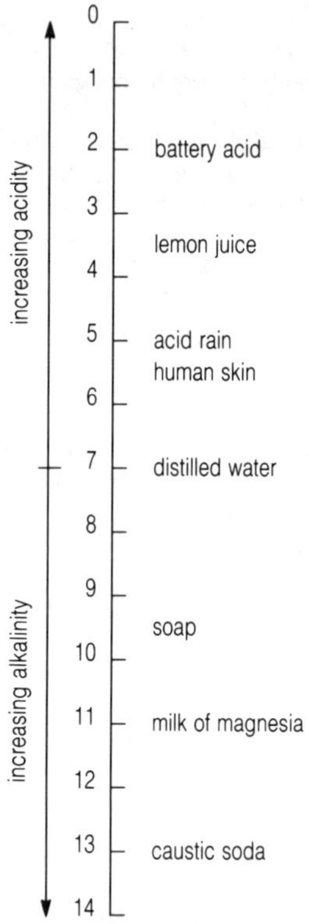

phenol member of a group of aromatic compounds with weakly acidic properties, which are characterized by a hydroxyl (-OH) group attached directly to an aromatic ring. The simplest of the phenols, derived from benzene, is also known as phenol and has the formula

phenol

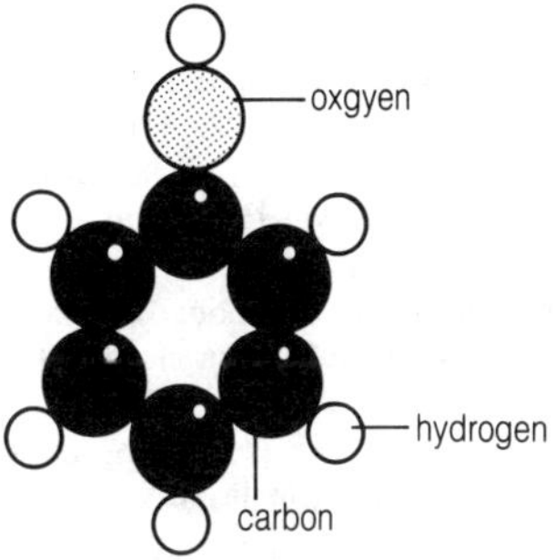

C_6H_5OH. It is also called ***carbolic acid*** and can be extracted from coal tar. Pure phenol consists of colourless, needle-shaped crystals which take up moisture from the atmosphere. It has a strong and characteristic smell and was once used as an antiseptic. It is, however, toxic by absorption through the skin.

phenolphthalein acid–base indicator that is clear below pH 8 and red above pH 9.6. It is used in titrating weak acids against strong bases.

phosphate salt or ester of ◊phosphoric acid. Incomplete neutralization of phosphoric acid gives rise to acid phosphates (see ◊acid salts and ◊buffer).

Phosphates are used as fertilizers, and lead to the development of healthy root systems. They are involved in many biochemical processes, often as part of complex molecules.

phosphoric acid acid derived from phosphorus and oxygen. Its commonest form (H_3PO_4) is also known as orthophosphoric acid, and is produced by the action of phosphorus pentoxide (P_2O_5) on water. It is used in rust removers and for rust-proofing iron and steel.

Its derivatives are part of many physiological and biochemical processes. A partly dehydrated form of phosphoric acid (metaphosphoric acid) exists as a glassy deliquescent polymeric solid with empirical formula HPO_3.

phosphorus (Greek *phosphoros* 'bearer of light') highly reactive, non-metallic element, symbol P, atomic number 15, relative atomic mass 30.9738. It occurs in nature as phosphates in the soil, in particular the mineral apatite, and is essential to both plant and animal life. The element has three allotropic forms: a black powder; a white-yellow, waxy solid that ignites spontaneously in air to form the poisonous gas phosphorous pentoxide; and a red-brown powder that neither ignites spontaneously nor is poisonous. Compounds of phosphorus are used in fertilizers, various organic chemicals, for matches and fireworks, and in glass and steel.

Phosphorus was first identified in 1669 by the German alchemist Hennig Brand, who prepared it from urine.

photochemical reaction any chemical reaction in which light is produced or light initiates the reaction. Light can initiate reactions by exciting atoms or molecules and making them more reactive: the light energy becomes converted to chemical energy. Many photochemical reactions set up a ◊chain reaction and produce ◊free radicals.

This type of reaction is seen in the bleaching of dyes or the yellowing of paper by sunlight. It is harnessed by plants in ◊photosynthesis and by humans in photography. Chemical reactions that produce light are most commonly seen when materials are burned. Light-emitting reactions are used by living organisms in bioluminescence. One photochemical reaction is the action of sunlight on car exhaust fumes, which results in the production of ◊ozone. Some large cities, such as Los Angeles, and Santiago, Chile, now suffer serious pollution due to photochemical smog.

photosynthesis process by which green plants trap light energy and use it to drive a series of chemical reactions, leading to the formation of ◊carbohydrates. All animals ultimately depend on photosynthesis because it is the method by which the basic food (sugar) is created. For photosynthesis to occur, the plant must possess ◊chlorophyll and must have a supply of carbon dioxide and water. Actively photosynthesizing green plants store excess sugar as starch (this can be tested for in the laboratory by using iodine).

The chemical reactions of photosynthesis occur in two stages. During the ***light reaction*** sunlight is used to split water (H_2O) into oxygen

(O_2), protons (hydrogen ions, H^+), and electrons, and oxygen is given off as a by-product. In the second-stage ***dark reaction***, for which sunlight is not required, the protons and electrons are used to convert carbon dioxide (CO_2) into carbohydrates ($C_m(H_2O)_n$). Photosynthesis depends on the ability of chlorophyll to capture the energy of sunlight and to use it to split water molecules.

pig iron or ***cast iron*** the crude, unrefined form of iron produced in a ◊blast furnace. It contains around 4% carbon plus some other impurities. ◊Cast iron is the partly refined form.

pipette device for the accurate measurement of a known volume of liquid, usually for transfer from one container to another, used in chemistry and biology laboratories.

A conventional pipette is a glass tube, often with an enlarged bulb, which is calibrated in one or more positions. Liquid is drawn into the pipette by suction, to the desired calibration mark. The release of liquid is controlled by careful pressure of the forefinger over the upper end of the tube, or by a plunger or rubber bulb.

pitch black, sticky substance, hard when cold, but liquid when hot, used for waterproofing, roofing, and paving. It is made by the destructive distillation of wood or coal tar, and has been used since antiquity in the waterproofing (caulking) of wooden ships.

pitchblende or ***uraninite*** brownish-black mineral, the major constituent of uranium ore, consisting mainly of uranium oxide (UO_2). It also contains some lead (the final, stable product of uranium decay) and variable amounts of most of the naturally occurring radioactive elements, which are products of either the decay or the fissioning of uranium isotopes. The uranium yield is 50–80%; it is also a source of radium, polonium, and actinium. Pitchblende was first studied by Pierre and Marie ◊Curie, who found radium and polonium in its residues in 1898.

plaster of Paris form of calcium sulphate, obtained from gypsum, mixed with water for making casts and moulds.

plastic any of the stable synthetic materials that are fluid at some stage in their manufacture, when they can be shaped, and that later set to rigid or semi-rigid solids. Plastics today are chiefly derived from petroleum. Most are polymers, made up of long chains of identical molecules mixed with additives, such as plasticizers, which improve flexibility, and antioxidants.

Processed by extrusion, injection-moulding, vacuum-forming and compression, they emerge in consistencies ranging from hard and inflexible to soft and rubbery. They replace an increasing number of natural substances, being lightweight, easy to clean, durable, and capable of being rendered very strong, for example by the addition of carbon fibres, for building aircraft and other engineering projects.

Thermoplastics soften when warmed, then re-harden as they cool. Examples of thermoplastics include polystyrene, a clear plastic used in kitchen utensils or (when expanded into a 'foam' by gas injection) in insulation and ceiling tiles; polyethylene or polythene, used for containers and wrapping; and polyvinyl chloride (PVC), used for drainpipes, floor tiles, audio discs, shoes, and handbags.

Thermosets remain rigid once set, and do not soften when warmed. They include bakelite, used in electrical insulation and telephone receivers; epoxy resins, used in paints and varnishes, to laminate wood, and as adhesives; polyesters, used in synthetic textile fibres and, with fibreglass reinforcement, in car bodies and boat hulls; and polyurethane, prepared in liquid form as a paint or varnish, and in foam form for upholstery and in lining materials (where it may be a fire hazard). One group of plastics, the silicones, are chemically inert, have good electrical properties, and repel water. Silicones find use in silicone rubber, paints, electrical insulation materials, laminates, waterproofing for walls, stain-resistant textiles, and cosmetics.

Shape-memory polymers are plastics that can be crumpled or flattened and will resume their original shape when heated. They include trans-polyisoprene and polynorbornene. The initial shape is determined by heating the polymer to over 35°C and pouring it into a metal mould. The shape can be altered with boiling water and the substance solidifies again when its temperature falls below 35°C.

Biodegradable plastics are increasingly in demand: Biopol was developed in 1990. The soil microbes involved build the plastic in their bodies from carbon dioxide and water (it constitutes 80% of their body). The unused parts of the microbe are dissolved away by heating in water. The discarded plastic can be placed in landfill sites where it breaks back down into carbon dioxide and water. It costs three to five times as much as ordinary plastics to produce.

platinum (Spanish *plata* 'little silver') heavy, soft, silver-white, malleable and ductile, metallic element, symbol Pt, atomic number 78, relative atomic mass 195.09. It is the first of a group of six metallic elements (platinum, osmium, iridium, rhodium, ruthenium, and palladium) that possess similar traits, such as resistance to tarnish, corrosion, and attack by acid, and that often occur as free metals (◊native metals). They often occur in natural alloys with each other, the commonest of which is osmiridium. Both pure and as an alloy, platinum is used in dentistry, jewellery, and as a catalyst.

plutonium silvery-white, radioactive, metallic element of the ◊actinide series, symbol Pu, atomic number 94, relative atomic mass 239.13. It occurs in nature in minute quantities in ◊pitchblende and other ores, but is produced in quantity only synthetically. It has six allotropic forms (see ◊allotropy) and is one of three fissile elements (elements capable of splitting into other elements—the others are thorium and uranium). The element has awkward physical properties and is the most toxic substance known.

Because Pu-239 is so easily synthesized from abundant uranium, it has been produced in large quantities by the weapons industry. It has a long half-life (24,000 years) during which time it remains highly toxic. Plutonium is dangerous to handle, difficult to store, and impossible to dispose of. It was first synthesized in 1940 by Glenn Seaborg and his team at the University of California at Berkeley, by bombarding uranium with neutrons.

pollution the harmful effect on the environment of by-products of human activity, principally industrial and agricultural processes—for example noise, smoke, car emissions, chemical effluents in seas and rivers, pesticides, sewage, and household waste. Pollution contributes to the ◊greenhouse effect.

Pollution control involves higher production costs for the industries concerned, but failure to implement adequate controls may result in irreversible environmental damage and an increase in the incidence of diseases such as cancer.

Natural disasters may also cause pollution; volcanic eruptions, for example, cause ash to be ejected into the atmosphere and deposited on land surfaces.

polychlorinated biphenyl (PCB) any of a group of chlorinated isomers of byphenal $(C_6H_5)_2$. They are dangerous industrial chemicals, valuable for their fire-resisting qualities. They constitute an environmental hazard because of their persistent toxicity. Since 1973 their use has been limited by international agreement.

polyester synthetic resin formed by the ◊condensation of polyhydric alcohols (alcohols containing more than one hydroxyl group) with dibasic acids (acids containing two replaceable hydrogen atoms). Polyesters are thermosetting ◊plastics, used in making synthetic fibres, such as Dacron and Terylene, and constructional plastics. With glass fibre added as reinforcement, polyesters are used in car bodies and boat hulls.

poly(ethene) or ***poly(ethylene)*** polymer of the gas ethene (ethylene) C_2H_4. It is a tough, white translucent waxy thermoplastic (which means it can be repeatedly softened by heating). It is used for packaging, bottles, toys, electric cable, pipes, and tubing.

Poly(ethene) is produced in two forms: low-density poly(ethene), made by high-pressure polymerization of ethene, and high-density poly(ethene), which is made at lower pressure by using catalysts. This form, first made by German chemist Karl Ziegler, is more rigid at low temperatures and softer at higher temperatures than the low-density type.

In the UK the plastic is better known under the trademark Polythene.

polymer compound made up of large, long-chain molecules composed of many repeated simple units (***monomers***). There are many polymers, both natural (cellulose, chitin, lignin) and synthetic (polyethylene and nylon, types of plastic). Synthetic polymers belong to two groups: thermosoftening and thermosetting (see ◊plastic).

polymerization the chemical union of two or more (usually small) molecules of the same kind to form a new compound.

addition polymerization In this type of polymerization, many molecules of a single compound join together to form long chains. An example is the polymerization of ethene to poly(ethene).

$$n\text{CH}_2\text{=CH}_2 \rightarrow [\text{CH}_2\text{CH}_2]_n$$

The only product of the reaction is the polymer. Other addition polymers include poly(vinyl chloride) (PVC).

condensation polymerization In this reaction, a small molecule such as water or hydrogen chloride is given off as a result of the polymerization. An example is the production of polyesters, formed by the polymerization of organic acids and alcohols. Condensation polymerization may involve a single monomer that has two reactive groups (such as an amino acid) or two or more different monomers (such as urea and formaldehyde, which polymerize to form resins).

polysaccharide long-chain ◊carbohydrate made up of hundreds or thousands of linked simple sugars (monosaccharides) such as glucose and closely related molecules.

The polysaccharides are natural polymers. They either act as energy-rich food stores in plants (starch) and animals (glycogen), or have structural roles in the plant cell wall (cellulose, pectin) or the tough outer skeleton of insects and similar creatures (chitin). See also ◊carbohydrate.

polystyrene polymer made from the polymer styrene (phenyl ethene, $C_2H_5CH{=}CH_2$); see ◊plastic.

poly(tetrafluoroethene) (PTFE) polymer made from the monomer tetrafluoroethene (CF_2CF_2). It is a thermosetting plastic with a high melting point that is used to produce 'non-stick' surfaces on pans and to coat bearings. Its trade name is Teflon.

Polythene trade name for a variety of ◊poly(ethene).

poly(urethane) any of a range of ◊plastics made by the polymerization of di-isocyanates with alcohols, polyesters, or polyethers.

polyvinyl chloride (PVC) polymer made from the monomer vinyl chloride (chloroethene, $CH_2{=}CHCl$); see ◊plastic.

potash general name for any potassium-containing mineral, most often applied to potassium carbonate (K_2CO_3) or potassium hydroxide (KOH). Potassium carbonate, originally made by roasting plants to ashes in earthenware pots, is commercially produced from the mineral sylvite (potassium chloride, KCl) and is used mainly in making artificial fertilizers, glass, and soap.

The potassium content of soils and fertilizers is also commonly expressed as potash, although in this case it usually refers to potassium oxide (K_2O).

potassium (Dutch *potassa* 'potash') soft, wax-like, silver-white, metallic element, symbol K (Latin *Kalium*), atomic number 19, relative atomic mass 39.0983. It is one of the ◊alkali metals and has a very low density—it floats on water, and is the second lightest metal (after lithium). It oxidizes rapidly when exposed to air and reacts violently with water. Of great abundance in the Earth's crust, it is widely distributed with other elements and found in salt and mineral deposits in the form of potassium aluminium silicates.

The element functions with sodium on the cellular level to make possible neuronal transmission and so is essential for animals; it is also essential for the growth of plants. It was discovered and named in 1807 by English chemist Humphry ◊Davy, who isolated it from potash in the first instance of a metal being isolated by electric current.

potassium dichromate $K_2Cr_2O_7$ orange, crystalline solid, soluble in water; it is a strong ◊oxidizing agent in the presence of dilute sulphuric acid. As it oxidizes other compounds it is itself reduced to potassium chromate (K_2CrO_4), which is green. Industrially it is used in the manufacture of dyes and glass and in tanning, photography, and ceramics.

potassium manganate(VII) or ***potassium permanganate*** $KMnO_4$ dark purple, crystalline solid, soluble in water; it is a strong ◊oxidizing agent in the presence of dilute sulphuric acid. In the process of oxidizing other compounds it is itself reduced to manganese(II) salts (containing the Mn^{2+} ion), which are colourless.

ppm abbreviation for ***parts per million***. An alternative (but numerically equivalent) unit used is milligrams per litre (mg l).

precipitate (ppt) solid (insoluble) substance sometimes produced when two solutions are mixed or when a gas is passed into a solution. In an equation the precipitate is often represented by the symbol (s).

$$AgNO_{3\,(aq)} + NaCl \rightarrow AgCl_{(s)} + NaNO_{3\,(aq)}$$

$$Ca(OH)_{2\,(aq)} + CO_{2\,(g)} \rightarrow CaCO_{3\,(s)} + H_2O_{(l)}$$

Priestley Joseph 1733–1804. English chemist who identified oxygen 1774.

A Unitarian minister, he was elected Fellow of the Royal Society 1766. In 1791 his chapel and house in Birmingham were sacked by a mob because of his support for the French Revolution. In 1794 he emigrated to the USA.

proof spirit numerical scale used to indicate the alcohol content of an alcoholic drink. Proof spirit (or 100% proof spirit) acquired its name from a solution of alcohol in water which, when used to moisten gunpowder, contained just enough alcohol to permit it to burn.

In practice, the degrees proof of an alcoholic drink is based on the specific gravity of an aqueous solution containing the same amount of alcohol as the drink. Typical values are: whisky, gin, and rum 70 degrees proof (40% alcohol); vodka 65 degrees proof; sherry 28 degrees proof; table wine 20 degrees proof; beer 4 degrees proof. The USA uses a different proof scale to the UK; a US whisky of 80 degrees proof on the US scale would be 70 degrees proof on the UK scale.

propane C_3H_8 gaseous hydrocarbon of the ◊alkane series, found in petroleum and used as fuel.

propanol common name ***propyl alcohol*** third member of the homologous series of ◊alcohols. Propanol is usually a mixture of two isomeric compounds (see ◊isomer): propan-1-ol ($CH_3CH_2CH_2OH$) and propan-2-ol ($CH_3CHOHCH_3$). Both are colourless liquids that can be mixed with water and are used in perfumery.

propanone common name ***acetone*** CH_3COCH_3 colourless inflammable liquid used extensively as a solvent, as in nail-varnish remover. It boils at 133.7°F/56.5°C, mixes with water in all proportions, and has a characteristic odour.

propene common name ***propylene*** $CH_3CH=CH_2$ second member of the ◊alkene series of hydrocarbons. A colourless, inflammable gas, it is widely used by industry to make organic chemicals, including polypropylene plastics.

properties the characteristics a substance possesses by virtue of its composition.

The ***physical properties*** of a substance can be measured by physical means, for example boiling point, melting point, hardness, elasticity, colour, and physical state. Its ***chemical properties*** are the ways in which it reacts with other substances: whether it is acidic or basic, an oxidizing or a reducing agent, a salt, or stable to heat, for example.

propyl alcohol common name for ◊propanol.

propylene common name for ◊propene.

protein long-chain molecule composed of amino acids joined by ◊peptide bonds. Proteins are essential to all living organisms. As ***enzymes*** they regulate all aspects of metabolism. Structural proteins such as ***keratin*** and ***collagen*** make up the skin, claws, bones, tendons, and ligaments; ***muscle*** proteins produce movement; ***haemoglobin*** transports oxygen; and ***membrane*** proteins regulate the movement of substances into and out of cells.

protein

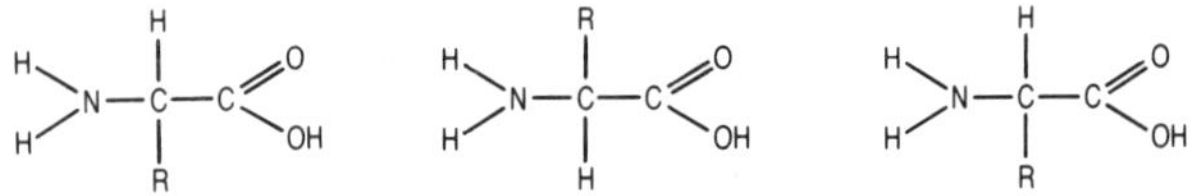

amino acids, where R is one of many possible side chains

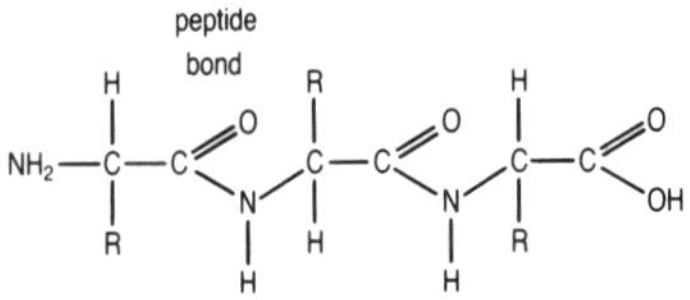

For humans, protein is an essential part of the diet, and is found in greatest quantity in soya beans and other grain legumes, meat, eggs, and cheese.

proton (Greek 'first') positively charged subatomic particle, a fundamental constituent of any atomic ◊nucleus. Its lifespan is effectively infinite.

A proton carries a unit positive charge equal to the negative charge of an ◊electron. Its mass is almost 1,836 times that of an electron, or 1.67×10^{-24} g. The number of protons in the atom of an ◊element is equal to its atomic, or proton, number.

proton number alternative name for ◊atomic number.

PTFE abbreviation for ◊***poly(tetrafluoroethene)***.

PVC abbreviation for ◊poly(vinylchloride).

pyrolysis decomposition of a substance by heating it to a high temperature in the absence of air.

Q

qualitative analysis procedure for determining the identity of the component(s) of a single substance or mixture. A series of simple reactions and tests can be carried out on a compound to determine the elements present.

quantitative analysis procedure for determining the precise amount of a known component present in a single substance or mixture. A known amount of the substance is subjected to particular procedures. Gravimetric analysis determines the mass of each constituent present; ◊volumetric analysis determines the concentration of a solution by ◊titration against a solution of known concentration.

quartz crystalline form of ◊silicon(IV) oxide (silica, SiO_2), one of the most abundant minerals of the Earth's crust (12% by volume). Quartz occurs in many different kinds of rock, including sandstone and granite. It is very hard, and is resistant to chemical and mechanical breakdown.

quasi-atom particle assemblage resembling an atom, in which particles not normally found in atoms become bound together for a brief period. Quasi-atoms are generally unstable structures, either because they are subject to matter–antimatter annihilation (positronium), or because one or more of their constituents is unstable (muonium).

quicklime common name for ◊calcium oxide.

quicksilver former name for the element ◊mercury.

R

radical group of atoms forming part of a molecule, which acts as a unit and takes part in chemical reactions without disintegrating, yet often cannot exist alone; for example, the methyl radical $-CH_3$; or the carboxylic acid radical $-COOH$.

radioactive decay spontaneous alteration of the nucleus of a radioactive atom; this changes its atomic number, thus transmuting one element into another, and is accompanied by the emission of radiation. Alpha and beta decay are the most common forms. In ***alpha decay*** (the loss of a helium nucleus – two protons and two neutrons) the atomic number decreases by two; in ***beta decay*** (the loss of an electron) the atomic number increases by one.

Less commonly occurring decay forms include heavy-ion emission, electron capture, and spontaneous fission (in each of these the atomic number decreases). The final product in all modes of decay is a stable element (see radioactivity).

radioactivity spontaneous alteration of the nuclei of radioactive atoms, accompanied by the emmission of radiation. It is the property exhibited by the radioactive ◊isotopes of stable elements and all isotopes of radioactive elements, and can be either natural or induced. See ◊radioactive decay.

Radioactivity establishes an equilibrium in parts of the nucleus of unstable radioactive substances, ultimately to form a stable arrangement of nucleons (protons and neutrons); that is, a non radioactive (stable) element. This is most frequently accomplished by the emission of ***alpha particles*** (helium nuclei); ***beta particles*** (electrons); or ***gamma rays*** (electromagnetic waves of very high frequency). It takes place either directly, through a one-step decay, or indirectly, through a number of decays that transmute one element into another. This is called a decay series or chain, and sometimes produces an element more reactive than its predecessor.

The instability of the particle arrangements in the nucleus of a radioactive atom (the ratio of neutrons to protons and/or the total number of both) determines the lengths of the ◊half-lives of the isotopes of that atom, which can range from fractions of a second to billions of years. Beta and gamma radiation are both ionizing and are therefore dangerous to body tissues, especially if a radioactive substance is ingested or inhaled.

radiochemistry study of radioactive isotopes and their compounds (whether produced from naturally radioactive or irradiated materials) and their use in the study of other chemical processes.

When such isotopes are used in labelled compounds, they enable the biochemical and physiological functioning of parts of the living body to be observed. They can help in the testing of new drugs, showing where the drug goes in the body and how long it stays there. They are also useful in diagnosis – for example, cancer, fetal abnormalities, and heart disease.

radium (Latin *radius* 'ray') white, radioactive, metallic element, symbol Ra, atomic number 88, relative atomic mass 226.02. It is one of the ◊alkaline-earth metals, found in nature in ◊pitchblende and other uranium ores. Of the 16 isotopes, the commonest, Ra-226, has a half-life of 1.622 years. The element was discovered and named in 1898 by Pierre and Marie ◊Curie, who were investigating the residues of pitchblende.

Radium decays in successive steps to produce radon (a gas), polonium, and finally a stable isotope of lead. The isotope Ra-223 decays through the uncommon mode of heavy-ion emission, giving off carbon-14 and transmuting directly to lead. As radium luminesces, it was formerly used in paints that glowed in the dark; when the hazards of radioactivity became known its use was abandoned, but factory and dump sites remain contaminated and many former workers and neighbours contracted fatal cancers.

radon colourless, odourless, gaseous, radioactive, non-metallic element, symbol Rn, atomic number 86, relative atomic mass 222. It is grouped with the ◊inert gases and was formerly considered non-reactive, but is now known to form some compounds with fluorine. Of the 20 known isotopes, only three occur in nature; the longest half-life is 3.82 days.

Radon is the densest gas known and occurs in small amounts in spring water, streams, and the air, being formed from the natural radioactive decay of radium. Ernest Rutherford discovered the isotope Rn-220 in 1899, and Friedrich Dorn (1848–1916) in 1900; after several other chemists discovered additional isotopes, William Ramsay isolated the element, which he named niton in 1908. The name radon was adopted in the 1920s.

rare-earth element or ***lanthanide*** any of a series of 15 metallic elements with atomic numbers 57 (lanthanum) to 71 (lutetium). One of its members, promethium, is radioactive. All occur in nature. Lanthanides are grouped because of their chemical similarities (they are all bivalent), their properties differing only slightly with atomic number.

rare gas alternative name for ◊inert gas.

rate of reaction the speed at which a chemical reaction proceeds. It is usually expressed in terms of the concentration (usually in ◊moles per litre) of a reactant consumed or product formed in unit time; so the units would be moles per litre per second (mol l^{-1} s^{-1}). The rate of a reaction is affected by the concentration of the reactants, the temperature of the reactants, and the presence of a ◊catalyst. If the reaction is entirely in the gas state, pressure affects the rate, and, for solids, the particle size.

During a reaction at constant temperature the concentration of the reactants decreases and so the rate of reaction decreases. These changes can be represented by drawing graphs.

The rate of reaction is at its greatest at the beginning of the reaction and it gradually slows down. Increasing the temperature produces large increases in the rate of reaction. A 10°C rise can double the rate while a 40°C rise can produce a 50–100-fold increase in the rate. ◊Collision theory is used to explain these effects. Increasing the concentration or the pressure of a gas means there are more particles per unit volume, therefore there are more collisions and more fruitful collisions. Increasing the temperature makes the particles move much faster, resulting in more collisions per unit time and more fruitful collisions; consequently the rate increases.

rate of reaction

(a) rate of reaction decreases with time

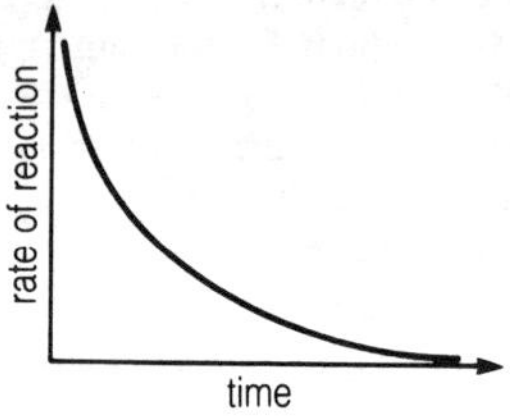

(b) concentration of reactant decreases with time

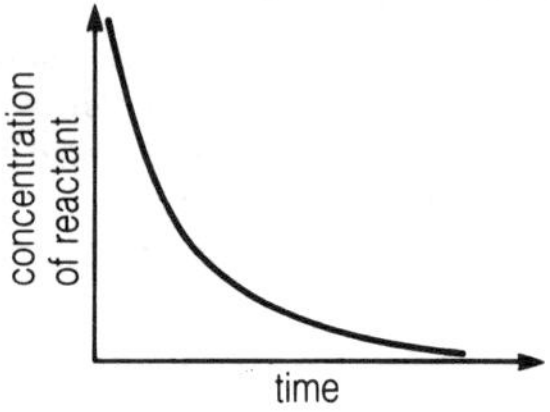

(c) concentration of product increases with time

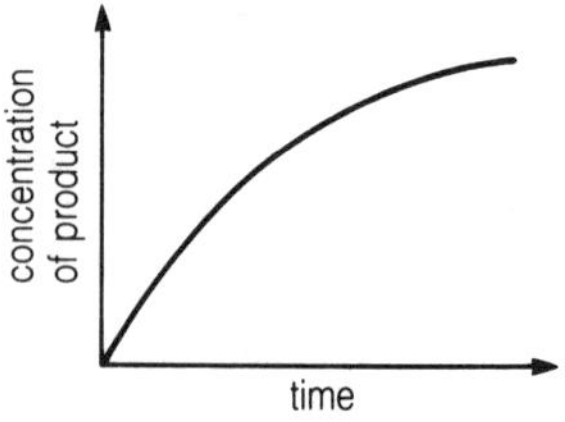

reaction the coming together of two or more atoms, ions or molecules resulting in a ◊chemical change. The nature of the reaction is portrayed by a chemical equation.

reactivity series series produced by arranging the metals in order of their ease of reaction with reagents such as oxygen, water, and acids. This arrangement aids in understanding the properties of metals, helps to explain differences, and enables predictions to be made about a certain metal based on a knowledge of its position or properties.

recycling processing of industrial and household waste (such as paper, glass, and some metals) so that it can be reused, thus saving expenditure on scarce raw materials, slowing down the depletion of non-renewable resources, and helping to reduce pollution.

redox reaction chemical change where one reactant is reduced and the other reactant oxidized. The reaction can only occur if both are present and each changes simultaneously. For example, hydrogen reduces copper(II) oxide to copper while it is itself oxidized to water.

$$CuO + H_2 \rightarrow Cu + H_2O$$

Many chemical changes can be classified as redox. Corrosion of iron, the reactions in chemical cells, and electrolysis are just a few instances where redox reactions occur.

reducing agent substance that reduces another substance (see ◊reduction). In a redox reaction, the reducing agent is the substance that is itself oxidized. Strong reducing agents include hydrogen, carbon monoxide, carbon, and metals.

reduction reaction in which an atom, ion, or molecule loses oxygen, gains hydrogen, or gains an electron. Examples include the reduction of iron(III) oxide to iron by carbon monoxide, the hydrogenation of ethene to ethane, and the reduction of a sodium ion to sodium.

$$Fe_2O_3 + 3CO \rightarrow 2Fe + 3CO_2$$

$$CH_2{=}CH_2 + H_2 \rightarrow CH_3{-}CH_3$$

$$Na^+ + e^- \rightarrow Na$$

reactivity series

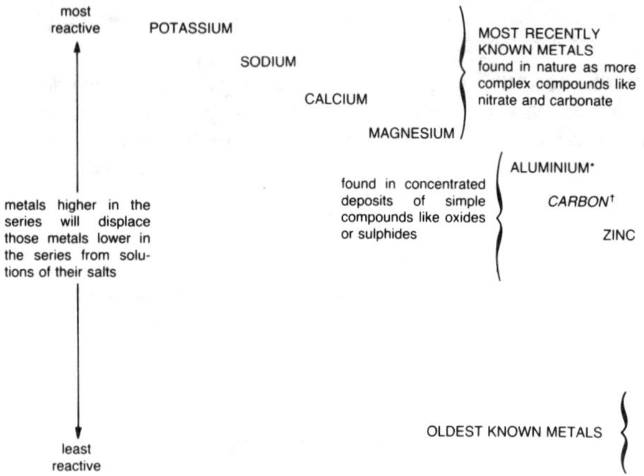

*The position of aluminium may appear anomalous as the oxide on its surface makes it unreactive.

relative atomic mass the mass of an atom. It depends on the number of protons and neutrons in the atom, the electrons having negligible mass. It is calculated relative to one-twelfth the mass of an atom of carbon-12. If more than one ◊isotope of the element is present, the relative atomic mass is calculated by taking an average that takes account of the relative proportions of each isotope, resulting in values that are not whole numbers. The term ***atomic weight***, although commonly used, is strictly speaking incorrect.

relative molecular mass the mass of a molecule, calculated relative to one-twelfth the mass of an atom of carbon-12. It is found by adding the relative atomic masses of the atoms that make up the molecule. The term ***molecular weight*** is often used, but is strictly incorrect.

residue substance or mixture of substances remaining in the original container after the removal of one or more components by a separation process.

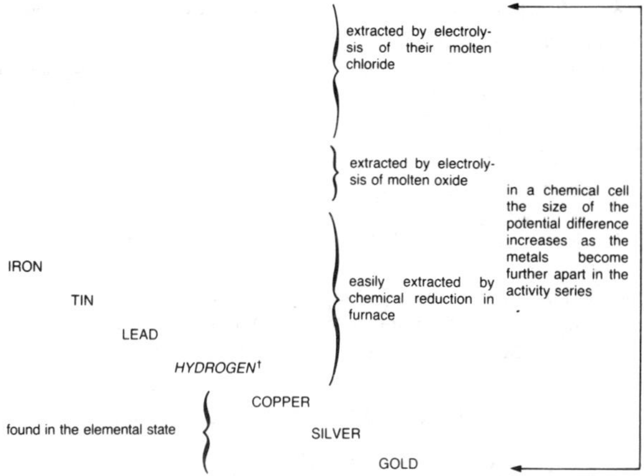

†Carbon and hydrogen, although non-metals, are included to show that they will only reduce metals below them in the reactivity series.

The non-volatile substance left in a container after the ◊evaporation of a liquid, the solid left behind after the removal of liquid by filtration, and the substances left in a distillation flask after the removal of components by ◊distillation, are all residues.

resin substance exuded from pines, firs, and other trees in gummy drops that harden in air. Varnishes are common products of the hard resins, and ointments come from the soft resins.

Rosin is the solid residue of distilled turpentine, a soft resin. The name 'resin' is also given to many synthetic products manufactured by polymerization; they are used in adhesives, plastics, and varnishes.

respiration biochemical process whereby food molecules are progressively broken down (oxidized) to release energy in the form of ATP. In most organisms this is requires oxygen, but in some bacteria the oxidant is the nitrate or sulphate ion instead.

reversible reaction chemical reaction that proceeds in both directions at the same time, as the product decomposes back into reactants as it is being produced. Such reactions do not run to completion, providing no substance leaves the system. Examples include the manufacture of ammonia from hydrogen and nitrogen, and the oxidation of sulphur dioxide to sulphur trioxide.

$$N_2 + 3H_2 \leftrightarrow 2NH_3$$

$$2SO_2 + O_2 \leftrightarrow 2SO_3$$

The term is also applied to those reactions that can be made to go in the opposite direction by changing the conditions, but these run to completion because some of the substances escape from the reaction. Examples are the decomposition of calcium hydrogencarbonate on heating and the loss of water of crystallization by copper(II) sulphate pentahydrate.

$$Ca(HCO_3)_{2\,(aq)} \rightarrow CaCO_{3\,(s)} + CO_{2\,(g)} + H_2O_{(l)}$$

$$CuSO_4.5H_2O_{(s)} \rightarrow CuSO_{4\,(s)} + 5H_2O_{(g)}$$

rhombic sulphur allotropic form of sulphur. At room temperature, it is the stable allotrope (see ◊allotropy), unlike ◊monoclinic sulphur.

rubidium (Latin *rubidus* 'red') soft, silver-white, metallic element, symbol Rb, atomic number 37, relative atomic mass 85.47. It is one of the ◊alkali metals, ignites spontaneously in air, and reacts violently with water. It is used in photoelectric cells and vacuum-tube filaments.

Rubidium was discovered spectroscopically by Robert Bunsen and Gustav Kirchhoff in 1861, and named after the red lines in its spectrum.

rust reddish-brown oxide of iron formed by the action of moisture and oxygen on the metal. It consists mainly of hydrated iron(III) oxide ($Fe_2O_3.H_2O$ and iron(III) hydroxide ($Fe(OH)_3$).

Paints that penetrate beneath any moisture, and plastic compounds that combine with existing rust to form a protective coating, are used to avoid rusting.

S

sal ammoniac former name for ◊ammonium chloride.

salicylic acid HOC_6H_4COOH an intermediate in the preparation of acetylsalicylic acid, the active chemical constituent of aspirin. The acid and its salts (salicylates) occur naturally in many plants; concentrated sources include willow bark and oil of wintergreen.

When purified, salicylic acid is a white solid that crystallizes into prismatic needles at 318°F/159°C. It is used as an antiseptic and in food preparation and dyestuffs as well as in the preparation of aspirin.

salt any member of a group of compounds containing a positive ion (cation) derived from a ◊metal or ammonia and a negative ion (anion) derived from an ◊acid or ◊non-metal. If the negative ion has a replaceable hydrogen atom it is an ***acid salt*** (for example, sodium hydrogensulphate, $NaHSO_4$; potassium phosphonate, KH_2PO_4; sodium hydrogencarbonate, $NaHCO_3$); if not, it is classed as a ***normal salt*** (for example sodium chloride, NaCl; potassium sulphate, K_2SO_4; magnesium nitrate, $Mg(NO_3)_2$). ***Common salt*** is sodium chloride. Salts have the properties typical of ionic compounds.

formula As all salts are electrically neutral, their formulae can be worked out by making sure that the total numbers of positive and negative charges arising from the ions are equal.

preparation Various methods can be used to prepare salts in the laboratory; the choice is dictated by the starting materials available and by whether the required salt is soluble or insoluble. Methods include:

(i) ***acid + metal*** for salts of magnesium, iron, and zinc;

(ii) ***acid + base*** for salts of magnesium, iron, zinc, and calcium;

(iii) ***acid + carbonate*** for salts of all metals;

(iv) ***acid + alkali*** for salts of sodium, potassium, and ammonium;

salt

	negative ions		
	NO_3^-	SO_4^{2-}	PO_4^{3-}
positive ions Na^+	$NaNO_3$ sodium nitrate	Na_2SO_4 sodium sulphate	Na_3PO_4 sodium phosphate
Mg^{2+}	$Mg(NO_3)_2$ magnesium nitrate	$MgSO_4$ magnesium sulphate	$Mg_3(PO_4)_2$ magnesium phosphate
Fe^{3+}	$Fe(NO_3)_3$ iron (III) nitrate	$Fe_2(SO_4)_3$ iron (III) sulphate	$FePO_4$ iron (III) phosphate

(v) ***direct combination*** for sulphides and chlorides;

(vi) ***double decomposition*** for insoluble salts.

In methods (i)–(iii) an excess of the solid reactant is added to the acid to ensure that no acid remains. The excess solid is filtered from the salt solution and the filtrate is boiled to a much smaller volume; it is then allowed to cool and crystallize. The salt crystals are filtered and dried on filter paper.

In method (iv) an indicator is used to determine the volume of acid needed to neutralize the alkali (or vice versa). The colour can then be removed by charcoal treatment, or alternatively the experiment can be repeated without the indicator. The solution is boiled to a smaller volume, cooled to crystallize the salt, and the crystals filtered and dried as in (i)–(iii) above.

In method (v) the salt is made in one step and does not require drying.

In method (vi) the two solutions are mixed and stirred. The precipitated salt is filtered, washed well with water to remove the soluble impurities, and allowed to dry in air or an oven at 60–80°C.

salt, common popular name for ◊sodium chloride (NaCl).

saltpetre former name for potassium nitrate (KNO_3), the compound used in making gunpowder (from about 1500). It occurs naturally, being deposited during dry periods in regions, such as India, possessing warm climates.

sal volatile another name for ◊smelling salts.

saponification the ◊hydrolysis (splitting) of an ◊ester by treatment with a strong alkali, resulting in the liberation of the alcohol from which the ester had been derived and a salt of the constituent fatty acid. The process is used in the manufacture of soap.

saturated compound organic compound that contains only single covalent bonds, such as propane. Saturated organic compounds can only undergo further reaction by ◊substitution reactions such as the production of chloropropane from propane. They have fewer reactions than unsaturated compounds such as alkenes.

$$C_3H_8 + Cl_2 \rightarrow C_3H_7Cl + HCl$$

saturated solution solution obtained when a solvent (liquid) can dissolve no more of a solute (usually a solid) at a particular temperature. Normally, a slight fall in temperature causes some of the solute to crystallize out of solution. If this does not happen the phenomenon is called supercooling, and the solution is said to be ***supersaturated***.

scale ◊calcium carbonate deposits that form on the inside of a kettle or boiler as a result of boiling ◊hard water.

Scheele Karl 1742–1786. Swedish chemist and pharmacist. In his book *Experiments on Air and Fire* 1777, he argued that the atmosphere was composed of two gases. One, which supported combustion (oxygen), he called 'fire air', and the other, which inhibited combustion (nitrogen), he called 'vitiated air'. He thus anticipated Joseph ◊Priestley's discovery of oxygen by two years.

sea water the water of the seas and oceans, covering about 70% of the Earth's surface and comprising about 97% of the world's water (only about 3% is fresh water). Sea water contains a large amount of dissolved solids, the most abundant of which is sodium chloride (almost 3% by mass); other salts include potassium chloride, bromide, and iodide, magnesium chloride, and magnesium sulphate. It also contains a large amount of dissolved carbon dioxide, and thus acts as a carbon 'sink' that may help to reduce the greenhouse effect.

selenium (Greek *Selene* 'Moon') grey, non-metallic element, symbol Se, atomic number 34, relative atomic mass 78.96. It belongs to the sulphur group and occurs in several allotropic forms that differ in their physical and chemical properties. It is an essential trace element in human nutrition. Obtained from many sulphide ores and selenides, it is used as a red colouring for glass and enamel.

Because its electrical conductivity varies with the intensity of light, selenium is used extensively in photoelectric devices. It was discovered in 1817 by Swedish chemist Jöns Berzelius and named after the Moon because its properties follow those of tellurium, whose name derives from the Latin *Tellus* 'Earth'.

semiconductor crystalline material with an electrical conductivity between that of metals (good) and insulators (poor).

The conductivity of semiconductors can usually be improved by minute additions of different substances or by other factors. Silicon, for example, has poor conductivity at low temperatures, but this is improved by the application of light, heat, or voltage; hence silicon is used in transistors, rectifiers, and integrated circuits (silicon chips).

silica common name for ◊silicon(IV) oxide SiO_2.

silicate compound containing silicon and oxygen combined together as a negative ion (◊anion), together with one or more metal ◊cations.

Common natural silicates are sands (common sand is the oxide of silicon known as silica). Glass is a manufactured complex polysilicate material in which other elements (boron in borosilicate glass) have been incorporated.

silicon (Latin *silicium* 'silica') brittle, non-metallic element, symbol Si, atomic number 14, relative atomic mass 28.086. It is the second most abundant element (after oxygen) in the Earth's crust and occurs in amorphous and crystalline forms. In nature it is found only in combination with other elements, chiefly with oxygen in silicates and silica (◊silicon(IV) oxide, SiO_2), the main constituents of sands and gravels.

Pottery glazes and glassmaking are based on the use of silica sands and date back into prehistory. Today the crystalline form of silicon is used as a deoxidizing and hardening agent in steel, and has become the basis of the electronics industry because of its ◊semiconductor properties, being used to make silicon chips (integrated circuits) for microprocessors.

The element was isolated by Swedish chemist Jöns Berzelius in 1823, having been named in 1817 by Scottish chemist Thomas Thomson by analogy with boron and carbon because of its chemical resemblance to these elements.

silicon(IV) oxide or ***silicon dioxide*** SiO_2 colourless or white solid, insoluble in water, that occurs naturally in various forms, the most familiar and pure of which is ◊quartz. It has a ◊giant molecular structure.

silver white, lustrous, extremely malleable and ductile, metallic element, symbol Ag (from Latin *argentum*), atomic number 47, relative atomic mass 107.868. It occurs in nature in ores and as a free metal; the chief ores are sulphides, from which the metal is extracted by smelting with lead. It is the best metallic conductor of both heat and electricity; its most useful compounds are the chloride and bromide, which darken on exposure to light and are the basis of photographic emulsions.

Silver is used ornamentally, for jewellery and tableware, for coinage, in eletroplating, electrical contacts, and dentistry, and as a solder.

slag the molten mass of impurities that is produced in the smelting or refining of metals.

The slag produced in the manufacture of iron in a ◊blast furnace floats on the surface above the molten iron. It contains mostly silicates, phosphates, and sulphates of calcium. When cooled, the solid is broken up and used as a core material in the foundations of roads and buildings.

slaked lime or ***calcium hydroxide*** $Ca(OH)_2$ substance produced by adding water to lime (calcium oxide, CaO). Much heat is given out and the solid crumbles as it absorbs water. A solution of slaked lime is called ◊limewater.

smelling salts or ***sal volatile*** a mixture of ammonium carbonate, bicarbonate, and carbamate together with other strong-smelling substances, formerly used as a restorative for dizziness or fainting.

smelting processing a metallic ore in a furnace to produce the metal. Oxide ores such as iron ore are smelted with coke (carbon), which reduces the ore into metal and also provides fuel for the process.

A substance such as ◊limestone is often added during smelting to facilitate the melting process and to form a slag, which dissolves many of the impurities present.

smokeless fuel fuel that does not give off any smoke when burnt, as all the carbon is fully oxidized to carbon dioxide (CO_2). Natural gas, oil, and coke are smokeless fuels.

soap mixture of the sodium salts of various ◊fatty acids: palmitic, stearic, and oleic. It is made by the action of sodium hydroxide (caustic soda) or potassium hydroxide (caustic potash) on fats of animal or vegetable origin. Soap makes grease and dirt disperse in water in a similar manner to a ◊detergent.

soda ash former name for ◊sodium carbonate (Na_2CO_3).

soda lime powdery mixture of calcium hydroxide and sodium hydroxide or potassium hydroxide, used in medicine and as a drying agent.

sodium soft, wax-like, silver-white, metallic element, symbol Na (from Latin *natrium*), atomic number 11, relative atomic mass 22.898. It is one of the ◊alkali metals and has a very low density, being light enough to float on water. It is the sixth most abundant element (the fourth most abundant metal) in the Earth's crust. Sodium is highly reactive, oxidizing rapidly when exposed to air and reacting violently with water. Its most familiar compound is sodium chloride (common salt), which occurs naturally in the oceans and in salt deposits left by dried-up ancient seas.

Other sodium compounds used industrially include sodium hydroxide (caustic soda, NaOH), sodium carbonate (washing soda, $Na_2CO_3.10H_2O$) and hydrogencarbonate (sodium bicarbonate, $NaHCO_3$), sodium nitrate (saltpetre, $NaNO_3$, used as a fertilizer), and sodium thiosulphate (hypo, $Na_2S_2O_3$, used as a photographic fixer). Thousands of tons of these are manufactured annually. Sodium metal is used to a limited extent in spectroscopy and in discharge lamps, and is alloyed with potassium as a heat-transfer medium in nuclear reactors. It was isolated from sodium hydroxide (caustic soda) in 1807 by Humphry ◊Davy.

sodium carbonate or ***soda ash*** Na_2CO_3 anhydrous white solid. The hydrated, crystalline form ($Na_2CO_3.10H_2O$) is also known as washing soda. It is made by the ◊Solvay process and used as a mild alkali, as it is hydrolysed in water.

$$CO^{2-}_{3\,(aq)} + H_2O_{(l)} \rightarrow HCO^{-}_{3\,(aq)} + OH^{-}_{(aq)}$$

It is used to neutralize acids, in glass manufacture, and in water softening.

sodium chloride or ***common salt*** NaCl white, crystalline compound found widely in nature. It is a a typical ionic solid with a high melting point (801°C); it is soluble in water, insoluble in organic solvents, and is a strong electrolyte when molten or in aqueous solution. It is found dissolved in sea water and as concentrated deposits of rock salt (halite). It is widely used in the food industry as a flavouring and preservative, and in the chemical industry in the manufacture of sodium, chlorine, and sodium carbonate.

While common salt is an essential part of our diet, some medical experts believe that excess salt, largely from processed food, can lead to high blood pressure and increased risk of heart attacks.

sodium hydrogencarbonate or ***bicarbonate of soda*** $NaHCO_3$ mild alkaline substance manufactured by the ◊Solvay process. It has the following chemical reactions.

with acids It behaves as an alkali in solution, neutralizing acids to form a salt, water, and carbon dioxide.

$$NaHCO_{3\,(aq)} + HCl_{(aq)} \rightarrow NaCl_{(aq)} + CO_{2\,(g)} + H_2O_{(l)}$$

with heat On heating it decomposes to give off carbon dioxide and water, leaving the carbonate behind.

$$2NaHCO_{3\,(s)} \rightarrow Na_2CO_{3\,(s)} + CO_{2\,(g)} + H_2O_{(l)}$$

with water It undergoes some hydrolysis in aqueous solution.

$$HCO^-_{3\,(aq)} + H_2O_{(l)} \rightarrow H_2CO_{3\,(aq)} + OH^-_{(aq)}$$

It is used as an ◊antacid in indigestion preparations. Other uses include baking powders and effervescent health drinks.

sodium hydroxide or ***caustic soda*** NaOH the commonest ◊alkali. The solid and the solution are corrosive. It is used to neutralize acids, and in the manufacture of soap and oven cleaners.

It is prepared industrially from sodium chloride by the ◊electrolysis of concentrated brine. The brine contains the ions Na^+, Cl^-, H^+, and OH^- in solution. Two of these ions are discharged at the electrodes. The reactions are as follows

negative electrode: $2H^+ + 2e^- \rightarrow H_2$

positive electrode: $2Cl^- - 2e^- \rightarrow Cl_2$

This means the Na+ and OH^- remain in solution, so it becomes sodium hydroxide (containing some sodium chloride). This solution is sometimes called caustic soda liquor and is about 40% NaOH.

soft water water that contains very few dissolved metal ions such as calcium (Ca^{2+}) or magnesium (Mg^{2+}). It lathers easily with soap, and no ◊scale is formed inside kettles or boilers. It has been found that the incidence of heart disease is higher in soft-water areas.

sol colloidal suspension of very small solid particles in a liquid that retains the physical properties of a liquid (see ◊colloid).

solid state of matter that holds its own shape (as opposed to a liquid, which takes up the shape of its container, or a gas, which totally fills its container). According to ◊kinetic theory, the atoms or molecules in a solid are not free to move but merely vibrate about fixed positions, such as those in crystal lattices.

solidification change of state of a substance from liquid or vapour to solid on cooling. It is the opposite of melting or sublimation.

The temperature at which a liquid starts to solidify is only the same as the temperature at which the solid starts to melt if it is a pure compound.

solubility measure of the amount of solute (usually a solid or gas) that will dissolve in a given amount of solvent (usually a liquid) at a particular temperature. Solubility may be expressed as grams of solute per 100 grams of solvent or, for a gas, in parts per million (ppm) of solvent.

solubility curve graph that indicates how the solubility of a substance varies with temperature. Most salts increase their solubility with an increase in temperature, as they dissolve endothermically. These curves can be used to predict which salt, and how much of it, will crystallize out of a mixture of salts.

solubility curve

comparative curves for copper(II) sulphate and potassium nitrate

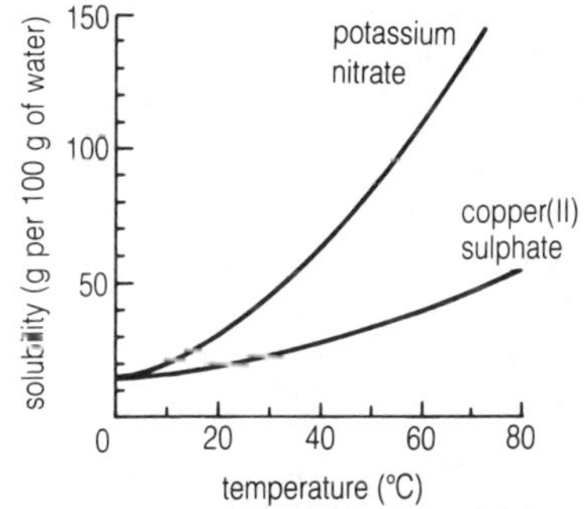

solute substance dissolved in another substance (see ◊solution).

solution two or more substances mixed to form a single, homogenous phase. One of the substances is the ***solvent*** and the others (***solutes***) are said to be dissolved in it.

The constituents of a solution may be solid, liquid, or gaseous. The solvent is normally the substance that is present in greatest quantity, although when one of them is a liquid this is considered to be the solvent even if it is not the major substance.

Solvay process industrial process for the manufacture of sodium carbonate.

It is a multistage process in which carbon dioxide is generated from limestone and passed through ◊brine saturated with ammonia. Sodium hydrogen carbonate is isolated and heated to yield sodium carbonate. All intermediate by-products are recycled so that the only ultimate by-product is calcium chloride.

It is named after its inventor, the Belgian chemist Ernest Solvay (1838–1922).

solvent substance, usually a liquid, that will dissolve another substance (see ◊solution). Although the commonest solvent is water, in popular use the term refers to low-boiling-point organic liquids that are harmful if used in a confined space. They can give rise to respiratory problems, liver damage, and neurological complaints.

Typical organic solvents are petroleum distillates (in glues), xylol (in paints), alcohols (for synthetic and natural resins such as shellac), esters (in lacquers, including nail varnish), ketones (in cellulose lacquers and resins), and chlorinated hydrocarbons (as paint stripper and dry-cleaning fluids). Some solvents have a pseudo-intoxicating effect which has given rise to the dangerous practice of glue-sniffing.

souring change that occurs to wine on prolonged exposure to air. The ethanol in the wine is oxidized by the air (oxygen) to ethanoic (acetic) acid. It is the presence of the ethanoic acid that produces the sour taste.

$$CH_3CH_2OH_{(aq)} + O_{2\,(g)} \rightarrow CH_3COOH_{(aq)} + H_2O_{(l)}$$

spectator ion in a chemical reaction that takes place in solution, an ion that remains in solution without taking part in the chemical change. For example, in the precipitation of barium sulphate from barium chloride and sodium sulphate, the sodium and chloride ions are spectator ions.

$$BaCl_{2\,(aq)} + Na_2SO_{4\,(aq)} \rightarrow BaSO_{4\,(s)} + 2NaCl_{(aq)}$$

speed of reaction alternative term for ◊rate of reaction.

standard temperature and pressure (STP) a standard set of conditions for experimental measurements, to enable comparisons to be made between sets of results. Standard temperature is 0°C and standard pressure 1 atmosphere (101,325 Pa).

starch widely distributed, high-molecular-mass ◊carbohydrate, produced by plants as a food store; its main dietary sources are cereals, legumes, and tubers, including potatoes. It consists of varying proportions of two ◊glucose polymers (◊polysaccharides): straight-chain (amylose) and branched (amylopectin) molecules.

Purified starch is a white powder used to stiffen textiles and paper and as a raw material for making various chemicals. It is used in the food industry as a thickening agent. Chemical treatment of starch gives rise to a range of 'modified starches' with varying properties. Hydrolysis (splitting) of starch by acid or enzymes generates a variety of 'glucose syrups' or 'liquid glucose' for use in the food industry. Complete hydrolysis of starch with acid generates the ◊monosaccharide glucose only. Incomplete hydrolysis or enzymic hydrolysis yields a mixture of glucose, maltose and non-hydrolysed fractions called 'dextrins'.

state change change in the physical state (solid, liquid, or gas) of a material. For instance, melting, boiling, evaporation, and their opposites (solidification and condensation) are state changes.

These changes require energy in the form of heat, called ◊latent heat, even though the temperature of the material does not change during the transition between states.

states of matter the forms (solid, liquid, or gas) in which material can exist. Whether a material is solid, liquid, or gas depends on its temperature and the pressure on it. The transition between states takes place at definite temperatures, called melting point and boiling point.

◊Kinetic theory describes how the state of a material depends on the movement and arrangement of its atoms or molecules. A hot ionized gas, or plasma, is often called the fourth state of matter, but liquid crystals, ◊colloids, and ◊glass also have a claim to this title.

state symbol symbol used in chemical equations to indicate the physical state of the substances present. The symbols are: (s) for solid, (l) for liquid, (g) for gas, and (aq) for aqueous.

steam dry, invisible gas formed by vaporizing water. The visible cloud that normally forms in the air when water is vaporized is due to minute suspended water particles. Steam is widely used in chemical and other industrial processes and for the generation of power.

stearic acid $CH_3(CH_2)_{16}COOH$ saturated long-chain ◊fatty acid, soluble in alcohol and ether but not in water. It is found in many fats and oils, and is used to make soap and candles and as a lubricant. The salts of stearic acid are called stearates.

steel alloy or mixture of iron and up to 1.7% carbon, sometimes with other elements, such as manganese, phosphorus, sulphur, and silicon. It has innumerable uses, including the manufacture of ships and cars, and the construction of tower-block frames and machinery of all kinds.

Steels with only small amounts of other metals are called ***mild*** or ***carbon steels***. These steels are far stronger than pure iron, with properties varying with the composition. ***Alloy steels*** include greater proportions of other metals; for example, stainless steel must contain at least 11% chromium.

Steel is produced by removing impurities, such as carbon, from raw or pig iron produced by a ◊blast furnace. The main industrial process is the ***basic-oxygen process***, in which pure oxygen is blown at high pressure through molten pig iron and scrap steel. The surface of the metal is disturbed by the blast and the oxidized impurities are burned out as gases or as slag. High-quality steel is made in an ***electric furnace***. A large electric current flows through electrodes in the furnace, melting a charge of scrap steel and iron. The quality of the steel produced can be controlled precisely because the temperature of the furnace can be maintained exactly and there are no combustion by-products to contaminate the steel.

strength of acids and bases the ability of ◊acids and ◊bases to dissociate in solution with water, and hence to produce a high or low ◊pH.

A strong acid is fully dissociated in aqueous solution, whereas a weak acid is only partly dissociated. Since the ◊dissociation of acids generates hydrogen ions, a solution of a strong acid will have a low pH. Similarly, a strong base will have a high pH, whereas a weaker base will not dissociate completely and will have a pH of nearer 7.

For a weak acid or base, the ease of dissociation, and hence strength, is given by its dissociation constant.

strontium soft, ductile, pale-yellow, metallic element, symbol Sr, atomic number 38, relative atomic mass 87.62. It is one of the ◊alkaline-earth elements, widely distributed in small quantities only as a sulphate or carbonate. Strontium salts burn with a red flame and are used in fireworks and signal flames.

The radioactive isotopes Sr-89 and Sr-90 (half-life 25 years) are some of the most dangerous products of the nuclear industry; they are fission products in nuclear explosions and in the reactors of nuclear power plants. Strontium is chemically similar to calcium and deposits in bones and other tissues, where the radioactivity is damaging. The element was named in 1808 by English chemist Humphry ◊Davy, who isolated it by electrolysis, after Strontian, a mining location in Scotland where it was first found.

sublimation the conversion of a solid to vapour without passing through the liquid phase.

Some substances that do not sublime at atmospheric pressure can be made to do so at low pressures. This is the principle of freeze-drying, during which ice sublimes at low pressure.

substitution reaction the replacement of one atom or ◊functional group in an organic molecule by another.

substrate in biochemistry, a compound or mixture of compounds acted on by an enzyme. The term also refers to a substance such as agar that provides the nutrients for the metabolism of microorganisms. Since the enzyme systems of microorganisms regulate their metabolism, the essential meaning is the same.

sucrose $C_{12}H_{22}O_{10}$ ◊disaccharide known commonly as ◊sugar, cane sugar, or beet sugar.

A single molecule of sucrose consists of a ◊glucose molecule and a fructose molecule bonded together. Sucrose does not have the reducing properties associated with most simple carbohydrates.

sugar sweet, soluble ◊carbohydrate, either a monosaccharide or a disaccharide. The major sources are tropical cane sugar, which accounts for about two-thirds of production, and temperate sugar beet. Polysaccharides, such as starch and cellulose, hydrolyse to many simple sugars.

sulphate SO_4^{2-} salt or ester derived from sulphuric acid. Most sulphates are water soluble (the exceptions are lead, calcium, strontium, and barium sulphates), and require a very high temperature to decompose them.

The commonest sulphates seen in the laboratory are copper(II) sulphate ($CuSO_4$), iron(II) sulphate ($FeSO_4$), and aluminium sulphate ($Al_2(SO_4)_3$). The ion is detected in solution by using barium chloride or barium nitrate to precipitate the insoluble sulphate.

sulphide compound of sulphur and another element in which sulphur is the more electronegative element (see ◊electronegativity). Sulphides occur in a number of minerals. Some of the more volatile sulphides have extremely unpleasant odours (hydrogen sulphide smells of bad eggs).

Pyrites (a naturally occurring sulphide of iron, FeS_2) is called 'fools' gold' because of its metallic yellow appearance.

sulphite SO_3^{2-} salt or ester derived from sulphurous acid.

sulphur brittle, pale-yellow, non-metallic element, symbol S, atomic number 16, relative atomic mass 32.064. It occurs in three allotropic forms: two crystalline (rhombic and monoclinic) and one amorphous. It burns in air with a blue flame and a stifling odour. Insoluble in water but soluble in carbon disulphide, it is a good electrical insulator.

sulphur

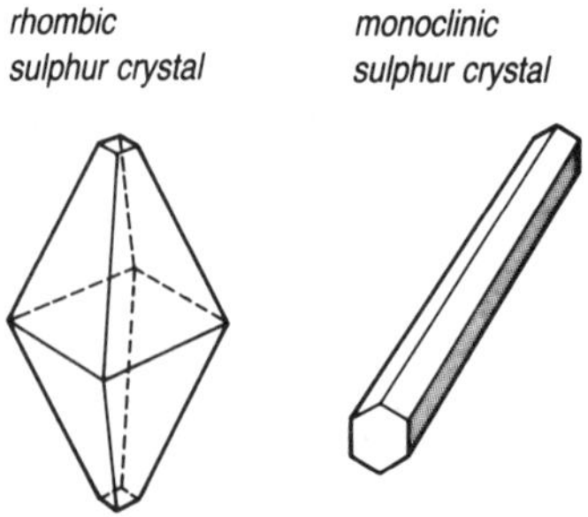

Sulphur is widely used in the manufacture of sulphuric acid (used to treat phosphate rock to make fertilizers) and in making paper, matches, gunpowder and fireworks; in vulcanizing rubber; and in medicines and insecticides.

It is found abundantly in nature in volcanic regions combined with both metals and non-metals, and also in its elemental form as a crystalline solid. It is a constituent of proteins, and has been known since ancient times.

sulphur dioxide SO_2 pungent gas produced by burning sulphur in air or oxygen. It is used widely for disinfecting food vessels and equipment, and as a preservative in some food products. It occurs in industrial flue gases and is a major cause of ◊acid rain.

sulphuric acid H_2SO_4 dense, viscous, colourless liquid that is extremely corrosive, causing severe burns. Its principal properties are as follows:

affinity for water Left open to the air, the concentrated acid increases its volume as it absorbs water vapour from the atmosphere. It is used to dry gases (except ammonia).

dehydration The concentrated acid is a powerful dehydrating agent that will remove the water of crystallization from hydrated salts such as copper sulphate crystals.

$$CuSO_4.5H_2O \leftrightarrow CuSO_4 + 5H_2O$$

It will also remove the elements of water from many organic compounds, for example sucrose.

$$C_{12}H_{22}O_{11} - 11H_2O \rightarrow 12C$$

oxidation Hot, concentrated sulphuric acid will oxidize metals and some non-metals.

$$Cu + 2H_2SO_4 \rightarrow CuSO_4 + SO_2 + 2H_2O$$

acidity Diluted, it acts as a strong, dibasic ◊acid, with the typical reactions of an acid.

$$H_2SO_4 + aq \leftrightarrow H^+_{(aq)} + HSO_4^-$$

The dilute acid should always be prepared by adding the concentrated acid to water with stirring and cooling.

On account of its chemical properties, sulphuric acid has many industrial uses. These include petrol refining and the manufacture of fertilizers, detergents, explosives, and dyes. In the UK more than two million tons of sulphuric acid are produced each year.

sulphurous acid H_2SO_3 solution of sulphur dioxide (SO_2) in water. It is a weak acid.

sulphur trioxide SO_3 colourless solid prepared by reacting sulphur dioxide and oxygen in the presence of a vanadium(V) oxide catalyst in the ◊contact process. It reacts violently with water to give sulphuric acid.

$$2SO_2 + O_2 \rightarrow 2SO_3$$

$$SO_3 + H_2O \rightarrow H_2SO_4$$

The violence of its reaction with water makes it extremely dangerous. In the contact process, it is dissolved in concentrated sulphuric acid to give oleum ($H_2S_2O_7$).

supersaturation the state of a solution that has a higher concentration of ◊solute than would normally be obtained in a ◊saturated solution.

Many solutes have a higher ◊solubility at high temperatures. If a hot saturated solution is cooled slowly, sometimes the excess solute does not come out of solution. This is an unstable situation and the introduction of a small solid particle will encourage the release of excess solute.

surgical spirit ◊ethanol to which has been added a small amount of methanol to render it unfit to drink. It may also contain other substances of clinical value such as oil of wintergreen, and is used to sterilize surfaces and to cleanse skin abrasions and sores.

suspension colloidal state consisting of small solid particles dispersed in a liquid or gas (see ◊colloid).

symbol letter or letters used to represent an element, usually derived from the beginning of its English or Latin name. Symbols derived from English include B, boron; C, carbon; Ba, barium; and Ca, calcium. Those derived from Latin include Na, sodium (Latin *natrium*); Pb, lead (Latin *plumbum*); and Au, gold (Latin *aurum*).

synthesis the formation of a substance or compound from more elementary compounds. The synthesis of a drug can involve several stages from the initial material to the final product; the complexity of these stages is a major factor in the cost of production.

T

tar dark brown or black viscous liquid obtained by the destructive distillation of coal, shale, and wood. Tars consist of a mixture of hydrocarbons, acids, and bases. ◊Creosote and ◊paraffin are produced from wood tar. See also ◊coal tar.

tartaric acid $HCOO(CHOH)_2COOH$ organic acid present in vegetable tissues and fruit juices in the form of salts of potassium, calcium, and magnesium. It is used in carbonated drinks and baking powders.

Teflon trade name for ◊poly(tetrafluoroethene) (PTFE), a tough, wax-like, heat-resistant plastic used for coating non-stick cookware and in gaskets and bearings.

tellurium (Latin *Tellus* 'Earth') silver-white, semi-metallic (◊metalloid) element, symbol Te, atomic number 52, relative atomic mass 127.60. Chemically it is similar to sulphur and selenium, and it is considered as one of the sulphur group. It occurs naturally in telluride minerals, and is used in colouring glass blue–brown, in the electrolytic refining of zinc, in electronics, and as a catalyst in refining petroleum.

It was discovered by the Austrian mineralogist Franz Müller (1740–1825) in 1782, and named in 1798 by the German chemist Martin Klaproth.

Terylene trade name for a polyester synthetic fibre produced by the chemicals company ICI. It is made by polymerizing ethane-1,2-diol (ethylene glycol) and benzene-1,4-dicarboxylic acid (terephthalic acid). Cloth made from Terylene keeps its shape after washing and is hard-wearing.

tetrachloromethane or ***carbon tetrachloride*** CCl_4 chlorinated organic compound that is a very efficient solvent for fats and greases. It is a toxic solvent and its use is restricted.

tetraethyl lead $Pb(C_2H_5)_4$ compound added to leaded petrol as a component of antiknock to increase the efficiency of combustion in car engines. It is a colourless liquid that is insoluble in water but soluble in organic solvents such as benzene, ethanol, and petrol.

thermal decomposition irreversible breakdown of a compound into simpler substances by heating it. The catalytic ◊cracking of hydrocarbons is an example.

thermal dissociation reversible breakdown of a compound into simpler substances by heating it (see ◊dissociation). The splitting of ammonium chloride into ammonia and hydrogen chloride is an example. On cooling, they recombine to form the salt.

$$NH_4Cl_{(s)} \leftrightarrow NH_{3\,(g)} + HCl_{(g)}$$

thermoplastic or ***thermosoftening plastic*** type of ◊plastic that always softens on repeated heating. Thermoplastics include polyethene, polystyrene, nylon, and polyester.

thermoplastic

monomer	*polymer*	*name*	*uses*
$CH_2{=}CH_2$ ethene	$\{CH_2\text{–}CH_2\}_n$	poly(ethene), polythene	bottles, packaging, insulation, pipes
$CH_2{=}CH\text{–}CH_3$ propene	$\{CH_2\text{–}CH\}_n$ | CH_3	poly(propene), polypropylene	mouldings, film, fibres
$CH_2{=}CH\text{–}Cl$ chloroethene (vinyl chloride)	$\{CH_2\text{–}CH\}_n$ | Cl	polyvinylchloride (PVC), poly(chloroethene)	insulation, flooring, household fabric
$CH_2{=}CH\text{–}C_6H_5$ phenylethene (styrene)	$\{CH_2\text{–}CH\}_n$ | C_6H_5	polystyrene, poly(phenylethene)	insulation, packaging
$CF_2{=}CF_2$ tetrafluoroethene	$\{CF_2\text{–}CF_2\}_n$ (n = 1000+)	poly(tetrafluoroethene) (PTFE)	high resistance to chemical and electrical reaction, low-friction applications

thermoset

monomer I	*monomer II*	*polymer name*	*uses*
formaldehyde (methanal)	phenol	PF resins (Bakelites)	electrical fittings, radio cabinets
formaldehyde	urea	UF resins	electrical fittings, insulation, adhesives
formaldehyde	melamine	melamines	laminates for furniture

thermoset or ***thermosetting plastic*** type of ◊plastic that remains rigid when set, and does not soften with heating. Thermosets have this property because the long-chain polymer molecules cross-link with each other to give a rigid structure. Examples include Bakelite, resins, melamine, and urea–formaldehyde resins.

tin soft, silver-white, malleable and somewhat ductile, metallic element, symbol Sn (from Latin *stannum*), atomic number 50, relative atomic mass 118.69. Tin exhibits ◊allotropy, having three forms: the familiar lustrous metallic form above 13.2°C/55.8°F; a brittle form above 161°C/321.8°F; and a grey powder form below 13.2°C/55.8°F . The metal is quite soft (slightly harder than lead) and can be rolled, pressed, or hammered into extremely thin sheets; it has a low melting point. In nature it occurs rarely as a free metal. It resists corrosion and is therefore used for coating and plating other metals.

Tin and copper smelted together form the oldest desired alloy, bronze; since the Bronze age (3,500 BC) that alloy has been the basis of both useful and decorative materials. Tin is also alloyed with metals other than copper to make solder and pewter. It was recognized as an element by Antoine Lavoisier, but the name is very old and comes from the Germanic form *zinn*.

titanium strong, light-weight, silver-grey, metallic element, symbol Ti, atomic number 22, relative atomic mass 47.90. The ninth most abundant element in the Earth's crust, its compounds occur in practically all igneous rocks and their sedimentary deposits. It is very strong and resistant to corrosion, and is used in building high-speed aircraft and spacecraft; it is also widely used in making alloys, as it unites with almost every metal except copper and aluminium.

The element was discovered in 1791 by English mineralogist William Gregor (1761–1817) and named in 1796 by German chemist Martin Klaproth after Titan, one of the giants of Greek mythology. It was not obtained in its pure form until 1925.

titration technique used to find the concentration of one compound in a solution by determining how much of it will react with a known amount of another compound in solution.

titration

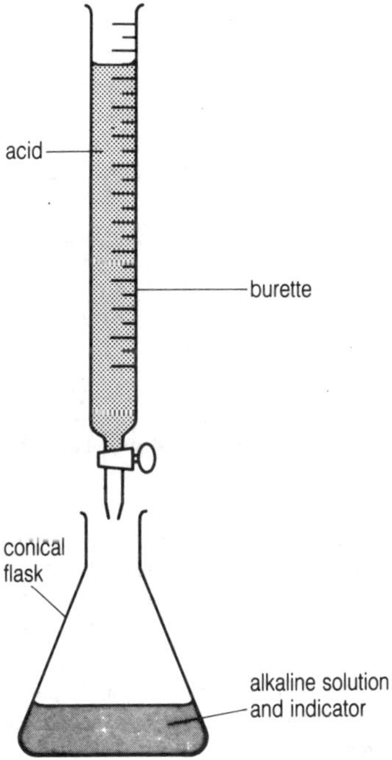

One of the solutions is measured by ◊pipette into the reaction vessel. The other is added a little at a time from a ◊burette. The end-point of the reaction is determined with an ◊indicator or an electrochemical device.

TNT abbreviation for ◊trinitrotoluene.

toluene or ***methyl benzene*** $C_6H_5CH_3$ colourless, inflammable liquid, insoluble in water, derived from petroleum. It is used as a solvent, in aircraft fuels, in preparing phenol (carbolic acid, used in making resins for adhesives, pharmaceuticals, and as a disinfectant), and the powerful high explosive ◊trinitrotoluene (TNT).

toxic poisonous or harmful. Lead from car exhausts, asbestos, and chlorinated solvents are some examples of toxic substances that occur in the environment; generally the effects take some time to become apparent (anything from a few hours to many years). The cumulative effects of toxic waste pose a serious threat to the ecological stability of the planet.

trace element element necessary in minute quantities for the health of a plant or animal. For example, magnesium, which occurs in chlorophyll, is essential to photosynthesis, and iodine is needed by the thyroid gland of mammals for making hormones that control growth and body chemistry.

transition metal any of a group of metallic elements that have incomplete inner electron shells and exhibit variable valency—for example, cobalt, copper, iron, and molybdenum. They are excellent conductors of electricity, and generally form highly coloured compounds.

trichloromethane or ***chloroform*** CCl_3 clear, colourless, toxic, carcinogenic liquid with a characteristic, pungent, sickly-sweet smell and taste, formerly used as an anaesthetic (now superseded by less harmful substances). It is used as a solvent and in the synthesis of organic chemical compounds.

triiodomethane or ***iodoform*** CHI_3 antiseptic that crystallizes into yellow hexagonal plates. It is soluble in ether, alcohol, and chloroform, but not in water.

trinitrotoluene (TNT) $CH_3C_6H_2(NO_2)_3$ powerful high explosive. It is a yellow solid, prepared in several isomeric forms from ◊toluene by using sulphuric and nitric acids.

tritium radioactive isotope of hydrogen, three times as heavy as ordinary hydrogen, consisting of one proton and two neutrons. It has a half-life of 12.5 years.

tungsten (Swedish *tung sten* 'heavy stone') hard, heavy, grey-white, metallic element, symbol W (from German *Wolfram*), atomic number 74, relative atomic mass 183.85. It occurs in the minerals wolframite, scheelite, and hubertite. It has the highest melting point of any metal (3,410°C/6,170°F) and is added to steel to make it harder, stronger, and more elastic; its other uses include high-speed cutting tools, electrical elements, and thermionic couplings. Its salts are used in the paint and tanning industries.

Tungsten was first recognized in 1781 by Swedish chemist Karl Scheele in the ore scheelite (originally called *tung sten* in Swedish). It was isolated in 1783 by the Spanish chemists Don Fausto D'Elhuyar (1755–1833) and his brother Juan José (1754–96).

U

universal indicator mixture of pH ◊indicators, each of which changes colour at a different pH value. The indicator is a different colour at different values of pH, ranging from red (at pH1) to purple (at pH13).

The pH of a substance may be found by adding a few drops of universal indicator and noting the colour, or by dipping an absorbent paper strip that has been impregnated with the indicator.

unleaded petrol petrol manufactured without the addition of antiknock. The use of unleaded petrol has been standard in the USA for some years, and is increasing in the UK. It has a slightly lower octane rating than leaded petrol, but has the advantage of not polluting the atmosphere with lead compounds. Many cars can be converted to run on unleaded petrol by altering the timing of the engine, and most new cars today are designed to do so. Cars fitted with a ◊catalytic converter must use unleaded fuel.

unsaturated compound compound in which two adjacent atoms are bonded by two or more covalent bonds.

Examples are ◊alkenes and ◊alkynes, where the two adjacent atoms are both carbon. The laboratory test for unsaturated compounds is to add bromine water, which is decolourized.

unsaturated solution solution that is capable of dissolving more solute than it already contains at the same temperature.

uranium hard, lustrous, silver-white, malleable and ductile, radioactive, metallic element of the ◊actinide series, symbol U, atomic number 92, relative atomic mass 238.029. It is the most abundant radioactive element in the Earth's crust, its decay giving rise to essentially all the radioactive elements in nature; its final decay product is the stable element lead. Uranium combines readily with most elements to form compounds that are extremely poisonous. The chief ore is ◊pitchblende, in which the element was discovered by German chemist Martin Klaproth in 1789; he named it after the planet Uranus, which had been discovered in 1781.

Uranium is one of three fissile elements (the others are thorium and plutonium). It was long considered to be the element with the highest atomic number to occur in nature. The isotopes U-238 and U-235 have been used to help determine the age of the Earth.

Uranium-238, which comprises about 99% of all naturally occuring uranium, has a half-life of 4.51×10^9 years. Because of its abundance, it is the isotope from which fissile plutonium is produced in breeder nuclear reactors. The fissile isotope U-235 has a half-life of 7.13×10^8 years and comprises about 0.7% of naturally occuring uranium; it is used directly as a fuel for nuclear reactors and in the manufacture of nuclear weapons.

urea $CO(NH_2)_2$ waste product formed in the mammalian liver when nitrogen compounds are broken down. It is excreted in urine. When purified, it is a white, crystalline solid. In industry it is used to make urea–formaldehyde plastics (or resins), pharmaceuticals, and fertilizers.

V

valence electron electron in the outermost shell of an atom. It is the valence electrons that are involved in the formation of ionic and covalent bonds (see ◊molecule). The number of electrons in this outermost shell represents the maximum possible ◊valency for many elements and matches the number of the group that the element occupies in the ◊periodic table of the elements.

valency the measure of an element's ability to combine with other elements, expressed as the number of atoms of hydrogen (or any other standard univalent element) capable of uniting with (or replacing) its atoms.

The elements are described as univalent, divalent, trivalent, and tetravalent when they unite with one, two, three, and four univalent atoms respectively. Some elements have ***variable valency***—for example, nitrogen and phosphorus can both possess valencies of either three or five. The valency of oxygen is two; hence the formula for water, H_2O (hydrogen being univalent).

valency shell the outermost shell of electrons in an atom. It contains the ◊valence electrons. Elements with four or more electrons in their outermost shell can show variable ◊valency. Chlorine can show valencies of 1, 3, 5, and 7 in different compounds.

valency shell

group number	*I*	*II*	*III*	*IV*	*V*	*VI*	*VII*
element	Na	Mg	Al	Si	P	S	Cl
atomic number	11	12	13	14	15	16	17
electron arrangement	2.8.1	2.8.2	2.8.3	2.8.4	2.8.5	2.8.6	2.8.7
valencies	1	2	3	4(2)	5(3)	6(2)	7(1)

vanadium silver-white, malleable and ductile, metallic element, symbol V, atomic number 23, relative atomic mass 50.942. It occurs in certain iron, lead, and uranium ores and is widely distributed in small quantities in igneous and sedimentary rocks. It is used to make steel alloys, to which it adds tensile strength.

The Spanish mineralogist Andrés del Rio (1764–1849) and the Swedish chemist Nils Sefström (1787–1845) discovered vanadium independently, the former in 1801 and the latter in 1831. Del Rio named it 'erythronium', but was persuaded by other chemists that he had not in fact discovered a new element; Sefström gave it its present name, after the Norse goddess of love and beauty, Vanadis (or Freya).

vanadium(V) oxide or ***vanadium pentoxide*** Va_2O_5 crystalline compound used as a catalyst in the ◊contact process for the manufacture of sulphuric acid.

van der Waals' force or ***intermolecular force*** weak force of attraction between molecules, caused by the electrons in the bonds between two atoms being attracted to the positive nucleus of another atom. The layers of carbon atoms in ◊graphite are held together by van der Waals' forces.

vaporization change of state of a substance from liquid to vapour. See ◊evaporation.

vapour one of the three states of matter (see also ◊solid and ◊liquid). The molecules in a vapour move randomly and are far apart, the distance between them, and therefore the volume of the vapour, being limited only by the walls of any vessel in which they might be contained. A vapour differs from a ◊gas only in that a vapour can be liquefied by increased pressure, whereas a gas cannot unless its temperature is lowered below a specific (critical) temperature; it then becomes a vapour and may be liquefied.

vinegar 4% solution of ethanoic (acetic) acid produced by the oxidation of alcohol, used to flavour food and as a preservative in pickles. ***Malt vinegar*** is brown and made from malted cereals; ***white vinegar*** is distilled from it. Other sources of vinegar include cider, inferior wine, and fermented honey.

vitriol any of a number of sulphate salts. Blue, green, and white vitriols are copper, ferrous, and zinc sulphate, respectively. ***Oil of vitriol*** is sulphuric acid.

volatile term describing a substance that readily passes from the liquid to the vapour phase.

volumetric analysis procedure used for determining the concentration of a solution. A known volume of a solution of unknown concentration is reacted with a solution of known concentration (standard). The standard solution is delivered from a ◊burette so the volume added is easily measured. This technique is known as ◊titration. Often an indicator is used to show when the correct proportions have reacted. This procedure is used for acid–base, ◊redox, and certain other reactions involving solutions.

W

washing soda or ***sodium carbonate decahydrate*** $Na_2CO_3.10H_2O$ substance added to washing water to 'soften' it (see ◊hard water).

water H_2O liquid without colour, taste, or odour, an oxide of hydrogen. Water is the most abundant substance on Earth, and is essential to all forms of life. It is a unique substance.

Water is a reactive substance; it reacts with many metals and non-metals as well as both inorganic and organic substances.

with metals Water reacts with many metals to give oxides or hydroxides. With sodium it reacts vigorously at room temperature; with zinc it forms zinc oxide when passed over it as steam at red heat.

$$2Na + 2H_2O \rightarrow 2NaOH + H_2$$

$$Zn + H_2O \rightarrow ZnO + H_2$$

with non-metals With chlorine, water forms hydrochloric acid and chloric(I) acid at room temperature; with carbon, it forms carbon monoxide and hydrogen ('water gas') when steam is passed over white-hot carbon.

$$Cl_2 + H_2O \rightarrow HCl + HOCl$$

$$C + H_2O \rightarrow CO + H_2$$

with inorganic compounds The most common reactions of water are hydrolysis (splitting) and hydration (adding water). Anhydrous copper sulphate is hydrated by the addition of water; sodium carbonate is hydrolysed to give sodium hydrogencarbonate and sodium hydroxide.

$$CuSO_4 + 5H_2O \rightarrow CuSO_4.5H_2O$$

$$Na_2CO_3 + H_2O \rightarrow NaHCO_3 + NaOH$$

with organic compounds Hydration (for example, the conversion of ethene to ethanol) and hydrolysis (as in the splitting of long-chain carbohydrates into smaller polysaccharides) are the commonest reactions.

$$CH_2{=}CH_2 + H_2O \rightarrow CH_3CH_2OH$$
$$C_{12}H_{22}O_{11} + H_2O \rightarrow 2C_6H_{12}O_6$$

The relative molecular mass of water is only 18, but its molecules are held together by intermolecular forces known as ◊hydrogen bonds. These arise between the oxygen atom of one water molecule and the hydrogen atom of an adjacent molecule, and help explain why water is a liquid even though its molecules are very small.

Water begins to freeze solid at 0°C/32°F, and to boil at 100°C/212°F. When liquid, it is virtually incompressible; frozen, it expands by 1/11 of its volume. At 4°C/39.2°F, one cubic centimetre of water has a mass of one gram, its maximum density, forming the unit of specific gravity. It has the highest known specific heat, and acts as an efficient solvent, particularly when hot. Most of the world's water is in the sea; less than 0.01% is fresh water.

water cycle the natural circulation of water through the biosphere. Water is lost from the Earth's surface to the atmosphere either by evaporation from the surface of lakes, rivers, and oceans or through the transpiration of plants. This atmospheric water forms clouds that condense to deposit moisture on the land and sea as rain or snow. The water that collects on land flows to the ocean in streams and rivers.

water gas fuel gas consisting of a mixture of carbon monoxide and hydrogen, made by passing steam over red-hot coke. The gas was once the chief source of hydrogen for chemical syntheses such as the ◊Haber process for making ammonia, but has been largely superseded in this and other reactions by hydrogen obtained from natural gas.

water glass or ***sodium metasilicate*** Na_2SiO_3 colourless, jelly-like substance that dissolves readily in water to give a solution used for preserving eggs and fireproofing porous materials such as cloth, paper, and wood. It is also used as an adhesive for paper and cardboard and in the manufacture of soap and silica gel, a substance that absorbs moisture.

water cycle

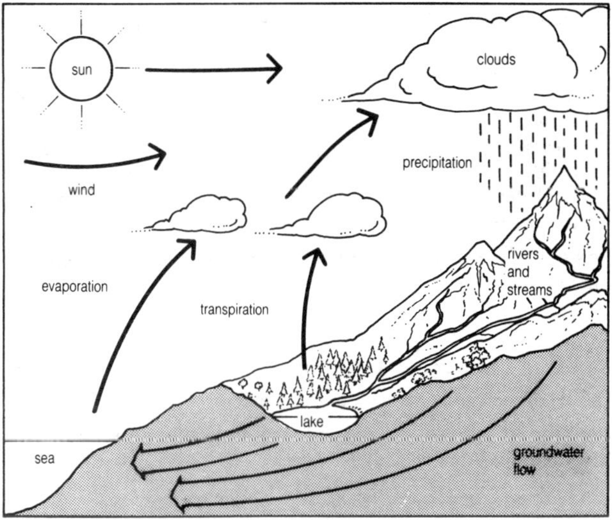

water of crystallization water chemically bonded to a salt in its crystalline state; for example, in copper(II) sulphate, there are five moles of water per mole of copper sulphate, hence its formula is $CuSO_4.5H_2O$. This water is responsible for the colour and shape of the crystalline form. When the crystals are heated gently, the water is driven off as steam and a white powder is formed.

$$CuSO_4.5H_2O_{(s)} \rightarrow CuSO_{4\,(s)} + 5H_2O_{(g)}$$

wax solid fatty substance of animal, vegetable, or mineral origin. Waxes are composed variously of ◊esters, ◊fatty acids, free ◊alcohols, and solid hydrocarbons.

Mineral waxes are obtained from petroleum and vary in hardness from the soft petroleum jelly (or petrolatum) used in ointments to the hard paraffin wax employed for making candles and waxed paper for drinks cartons.

Animal waxes include beeswax, the wool wax lanolin, and spermaceti from sperm whale oil; they are used mainly in cosmetics, ointments, and polishes. Another animal wax is tallow, a form of suet obtained from cattle and sheep's fat, once widely used to make candles and soap.

Vegetable waxes, which usually occur as a waterproof coating on plants that grow in hot, arid regions, include carnauba wax (from the leaves of the carnauba palm) and candelilla wax, both of which are components of hard polishes such as car waxes.

weak acid acid that only partially ionizes in aqueous solution (see ◊dissociation). Weak acids include ethanoic acid and carbonic acid.

$$CH_3COOH_{(l)} + aq \leftrightarrow 1\ H^+_{(aq)} + CH_3COO^-_{(aq)}$$

$$H_2CO_{3\ (aq)} \leftrightarrow H^+_{(aq)} + HCO^-_{3\ (aq)}$$

The pH of such acids lies between pH 3 and pH 6.

weak base base that only partially ionizes in aqueous solution (see ◊dissociation). Ammonia is a weak base.

$$NH_{3\ (g)} + H_2O_{(l)} \leftrightarrow NH^+_{4\ (aq)} + OH^-_{(aq)}$$

The pH of weak bases lies between pH 8 and pH 10.

weak electrolyte electrolyte that conducts electricity only moderately. Weak acids and bases are weak electrolytes.

wrought iron fairly pure iron containing some beads of ◊slag, widely used for construction work before the days of cheap steel. It is strong, tough, and easy to machine. It is made in a puddling furnace, invented by Henry Colt in England 1784. Pig iron is remelted and heated strongly in air with iron ore, burning out the carbon in the metal, leaving relatively pure iron and a slag containing impurities. The resulting pasty metal is then hammered to remove as much of the remaining slag as possible. It is still used in fences and grating.

X

xanthophyll yellow pigment in plants that, like ◊chlorophyll, is responsible for the production of carbohydrates by photosynthesis.

xenon (Greek *xenos* 'stranger') colourless, odourless, gaseous, non-metallic element, symbol Xe, atomic number 54, relative atomic mass 131.30. It is grouped with the ◊inert gases and was long believed not to enter into reactions, but is now known to form some compounds, mostly with fluorine. It is a heavy gas present in very small quantities in the air (about one part in 20 million).

Xenon is used in bubble chambers, light bulbs, vacuum tubes, and lasers. It was discovered in 1898 in a residue from liquid air by William Ramsay and Morris Travers (1872–1961).

Y

yeast one of various single-celled fungi (especially the genus *Saccharomyces*). When placed in a sugar solution the cells multiply and convert the sugar into ethanol and carbon dioxide. Yeasts are used as fermenting agents in baking, brewing, and the making of wine and spirits.

Z

zeolite any of the hydrous aluminium silicates, also containing sodium, calcium, barium, strontium, and potassium, chiefly found in igneous rocks and characterized by a ready loss or gain of water. Zeolites are used as 'molecular sieves' to separate mixtures because they are capable of selective absorption. They have a high ion-exchange capacity and can be used to make petrol, benzene, and toluene from low-grade raw materials, such as coal and methanol. Permutit is a synthetic zeolite used to soften hard water.

zinc (Germanic *zint* 'point') hard, brittle, bluish-white, metallic element, symbol Zn, atomic number 30, relative atomic mass 65.37. The principal ore is spalerite or zinc blende (zinc sulphide, ZnS). Zinc is little affected by air or moisture at ordinary temperatures; its chief uses are in alloys such as brass and in coating metals (for example, galvanized iron). Its compounds include zinc oxide, used in ointments (as an astringent), cosmetics, paints, glass, and printing ink.

Zinc has been used as a component of brass since the Bronze Age, but it was not recognized as a separate metal until 1746, when it was described by German chemist Andreas Sigismund Marggraf (1709–82). The name derives from the shape of the crystals on smelting.

zinc chloride $ZnCl_2$ white, crystalline compound that is deliquescent and sublimes easily. It is used as a catalyst, as a dehydrating agent, and as a flux in soldering.

zinc oxide ZnO white powder, yellow when hot, that occurs in nature as the mineral zincite. It is used in paints and as an antiseptic in zinc ointment; it is the main ingredient of calamine lotion.

zinc sulphide ZnS yellow-white solid that occurs in nature as the mineral sphalerite (also called zinc blende). It is the principal ore of zinc, and is used in the manufacture of fluorescent paints.

zwitterion ion that has both a positive and a negative charge, such as an ◊amino acid in neutral solution. For example, the amino acide glycine contains both a basic amino group (NH^2) and an acidic carboxyl group (-COOH); when both these are ionized in aqueous solution, the acid group loses a proton to the amino group, and the molecule is positively charged at one end and negatively charged at the other.

$$H_2NCH_2COOH \leftrightarrow {}^{+}H_3NCH_2COO^{-}$$

Appendix I

Periodic table of the elements

1
Hydrogen
H
1 00794

1 — atomic number
Hydrogen — name
H — symbol
1 00794 — relative atomic mass

element

1 Hydrogen **H** 1 00794								
3 Lithium **Li** 6941	4 Beryllium **Be** 901218							
11 Sodium **Na** 22 98977	12 Magnesium **Mg** 24 305							
19 Potassium **K** 39098	20 Calcium **Ca** 40 06	21 Scandium **Sc** 44 9559	22 Titanium **Ti** 4790	23 Vanadium **V** 50 9414	24 Chromium **Cr** 51 996	25 Manganese **Mn** 54 9380	26 Iron **Fe** 55 847	27 Cobalt **Co** 58 9332
37 Rubidium **Rb** 85 4678	38 Strontium **Sr** 8762	39 Yttrium **Y** 88 9059	40 Zirconium **Zr** 91 22	41 Niobium **Nb** 92 9064	42 Molybdenum **Mo** 9594	43 Technetium **Tc** 97 9072	44 Ruthenium **Ru** 101 07	45 Rhodium **Rh** 102 9055
55 Caesium **Cs** 132 9054	56 Barium **Ba** 137 34		72 Hafnium **Hf** 178 49	73 Tantalum **Ta** 180 9479	74 Tungsten **W** 183 85	75 Rhenium **Re** 186 207	76 Osmium **Os** 190 2	77 Iridium **Ir** 192 22
87 Francium **Fr** 223 0197	88 Radium **Ra** 226 0254		104 Unnilquadium **Unq** 216 109	105 Unnilpentium **Unp** 262 114	106 Unnilhexium **Unh** 263 120	107 Unnilseptium **Uns** 262	108 Unniloctium **Uno** 265	109 Unnilenium **Une** 266

Lanthanide series

57 Lanthanum **La** 138 9055	58 Cerium **Ce** 140 12	59 Praeseodymium **Pr** 140 9077	60 Neodymium **Nd** 144 24	61 Promethium **Pm** 144 9128	62 Samarium **Sm** 150 36

Actinide series

89 Actinium **Ac** 227 0278	90 Thorium **Th** 232 0381	91 Protactinium **Pa** 231 0359	92 Uranium **U** 238 029	93 Neptunium **Np** 237 0482	94 Plutonium **Pu** 244 0642

								2 Helium **He** 4 00260
			5 Boron **B** 10 81	6 Carbon **C** 12 011	7 Nitrogen **N** 14 0067	8 Oxygen **O** 15 9994	9 Fluorine **F** 18 99840	10 Neon **Ne** 20 179
			13 Aluminium **Al** 26 98154	14 Silicon **Si** 28 086	15 Phosphorus **P** 30 97376P	16 Sulphur **S** 32 06	17 Chlorine **Cl** 35 453	18 Argon **Ar** 39 948
28 Nickel **Ni** 58 70	29 Copper **Cu** 63 546	30 Zinc **Zn** 65 38	31 Gallium **Ga** 69 72	32 Germanium **Ge** 72 59	33 Arsenic **As** 74 9216	34 Selenium **Se** 78 96	35 Bromine **Br** 79 904	36 Krypton **Kr** 83 80
46 Palladium **Pd** 106 4	47 Silver **Ag** 107 868	48 Cadmium **Cd** 112 40	49 Indium **In** 114 82	50 Tin **Sn** 118 69	51 Antimony **Sb** 121 75	52 Tellurium **Te** 127 75	53 Iodine **I** 126 9045	54 Xenon **Xe** 131 30
78 Platinum **Pt** 195 09	79 Gold **Au** 196 9665	80 Mercury **Hg** 200 59	81 Thallium **Tl** 204 37	82 Lead **Pb** 207 37	83 Bismuth **Bi** 207 2	84 Polonium **Po** 210	85 Astatine **At** 211	86 Radon **Rn** 222 0176

63 Europium **Eu** 151 96	64 Gadolinium **Gd** 157 25	65 Terbium **Tb** 158 9254	66 Dysprosium **Dy** 162 50	67 Holmium **Ho** 164 9304	68 Erbium **Er** 167 26	69 Thulium **Tm** 168 9342	70 Ytterbium **Yb** 173 04	71 Lutetium **Lu** 174 97
95 Americium **Am** 243 0614	96 Curium **Cm** 247 0703	97 Berkelium **Bk** 247 0703	98 Californium **Cf** 251 0786	99 Einsteinium **Es** 252 0828	100 Fermium **Fm** 257 0951	101 Mendelevium **Md** 258.0986	012 Nobelium **No** 259 1009	103 Lawrencium **Lr** 260 1054

Appendix II

SI units and multiples

quantity	*SI unit*	*symbol*
absorbed radiation dose	gray	Gy
amount of substance	mole*	mol
electric capacitance	farad	F
electric charge	coulomb	C
electric conductance	siemens	S
electric current	ampere*	A
energy or work	joule	J
force	newton	N
frequency	hertz	Hz
illuminance	lux	lx
inductance	henry	H
length	metre*	m
luminous flux	lumen	lm
luminous intensity	candela*	cd
magnetic flux	weber	Wb
magnetic flux density	tesla	T
mass	kilogram*	kg
plane angle	radian	rad
potential difference	volt	V
power	watt	W
pressure	pascal	Pa
radiation dose equivalent	sievert	Sv
radiation exposure	roentgen	r
radioactivity	becquerel	Bq
resistance	ohm	W
solid angle	steradian	sr
sound intensity	decibel	dB
temperature	°Celsius	°C
temperature, thermodynamic	kelvin*	K
time	second*	s

*SI base unit

SI prefixes

multiple	*prefix*	*symbol*	*example*
1,000,000,000,000,000,000 (10^{18})	exa-	E	Eg(exagram)
1,000,000,000,000,000 (10^{15})	peta-	P	PJ (petajoule)
1,000,000,000,000 (10^{12})	tera-	T	TV (teravolt)
1,000,000,000 (10^{9})	giga-	G	GW (gigawatt)
1,000,000 (10^{6})	mega-	M	MHz (megahertz)
1,000 (10^{3})	kilo-	k	kg (kilogram)
100 (10^{2})	hecto-	h	hm (hectometre)
10	deca-	da-	daN (decanewton)
1/10 (10^{-1})	deci-	d	dC(decicoulomb)
1/100 (10^{-2})	centi-	c	cm(centimetre)
1/1,000 (10^{-3})	milli-	m	mA(milliampere)
1/1,000,000 (10^{-6})	micro-	m	µF (microfarad)
1/1,000,000,000 (10^{-9})	nano-	n	nm(nanometre)
1/1,000,000,000,000 (10^{-12})	pico-	p	ps (picosecond)
1/1,000,000,000,000,000 (10^{-15})	femto-	f	frad(femtoradian)
1/1,000,000,000,000,000,000 (10^{-18})	atto-	a	aT(attotesla)

The Hutchinson Pocket Encyclopedia 1994 edition

The Hutchinson Pocket Encyclopedia packs as much information into its 640 pages as other volumes costing half as much again. Written in clear, jargon-free language, it covers all areas of knowledge from science and engineering to biographies of major sporting, cultural, and political figures, to newly updated information on all the countries of the world.

With more than 10,000 up-to-date entries, 150 maps and diagrams, and a host of useful chronologies and tables, the Pocket is quite simply the most comprehensive compact encyclopedia available.

ISBN 0 09 178286 4